British Research and Developmental Aircraft

Volume 2

CRAIG VAUGHTON

Key Books

INNOVATION AND DISCOVERY SERIES, VOLUME 3

Front cover image: TSR.2, a byword for prematurely cancelled British aircraft projects.

Title page image: A fine study of the Supermarine Swift prototype WJ960, which first flew on 1 August 1951.

Contents page image: Westland's Widgeon was a development of its first licensed Sikorsky design, the Dragonfly, but saw limited success with only 15 produced. Ex-Bristow helicopters Widgeon G-AOZE is seen here at the Helicopter Museum.

Back cover image: XD350, Fairey's original FD.1 delta-winged research aircraft.

DEDICATION

To engineers and technicians worldwide, who have infinite patience for complex, contradictory and at times downright awkward machinery or electronics, yet believe that people should know better.

Published by Key Books
An imprint of Key Publishing Ltd
PO Box 100
Stamford
Lincs PE9 1XQ

www.keypublishing.com

ISBN 978 1 83632 236 8

Typeset by SJmagic DESIGN SERVICES, India.

Contents

Introduction

Of all the numerous research projects that took to the skies during the 1950s and '60s, as well as the aircraft for which an official requirement had been issued, yet were cancelled whilst under construction, the one that most aviation enthusiasts will know is TSR.2, which graces the front cover of this book. More words have probably been written about the might-have-beens surrounding this aircraft than any other type from any nation, let alone Britain.

TSR.2 was not unique. The completion of many government-instigated projects was curtailed due to plummeting finances, while on occasion, reduced faith in their own designs and a lack of company funds to continue a build in the hope of achieving export sales prevented a few promising private developments from emerging. As a result, next to nothing got off the ground unless there was an official government requirement and hence funding, followed by guaranteed sales. This often saw other countries design, build

Former Lightning Training Flight T.5 XS417 is at Newark Air Museum. Based on the F3, the T5 was fully operational and could accept the same *Firestreak* or *Red Top* missiles as the F3, but lacked the 30mm Aden cannon pack.

and then sell worldwide aircraft similar to projects that UK manufacturers had already cut metal for, or that had designs based on research aircraft that had flown. That this was repeated numerous times, under several different governments, seems unbelievable to the historian looking back.

Volume 1 covered the impact of the 1957 Defence White Paper, V-Bombers and the nuclear weapons they were built to carry, plus products from Hawker, Avro and Blackburn, as well as the series of aircraft that finally led to the Harrier. This volume deals with the other aircraft prototypes and development machines that were predominantly designed and built by the companies that merged to form BAC (the British Aircraft Corporation). It also includes the rotary-winged aircraft of Fairey, Bristol and Saunders-Roe (Saro), along with the first product released under the Westland banner once the other helicopter companies or divisions themselves had been merged into this single entity. There is a brief diversion into the world of air-to-air missiles, highlighting the impact that transistors had on this field and finally, there is a look at the work carried out by countless hard-working people behind the scenes, weaving engineering magic restoring aircraft to display condition.

Chapter 1

Rationalisation

Many may think that the business mergers we see today and the amalgamation of companies into larger entities began during the late 1960s, facilitated by increasing global communications, computerisation and jet travel. But look further back and you will find that some of the companies involved in the British aircraft business had grown in a similar manner, though on a more national scale, for more than a century. Vickers, which formed part of BAC, began as a Sheffield steel company in 1828, yet by 1928 had become one of the largest multi-faceted engineering companies in Britain and across what was still an empire, through a century of acquisitions and mergers. From guns and ships to tanks and heavy engineering, Vickers was involved in them all. The company's interests in aviation started in 1911 and for the most part, the Vickers name became associated with bombers and passenger aircraft, from the Vimy and Wellington through to the

Britain's first fully operational air-to-air missile, the de Havilland *Firestreak*, was developed under the code name of *Blue Jay* and was placed in service in 1957. One is seen slung underneath Newark Air Museum's de Havilland Sea Vixen XJ560, but was a common sight on Lightnings and Javelins through the 1960s and '70s.

Viscount and VC10. In 1928, Vickers had purchased Supermarine Aviation, which of course went on to develop the Schneider Trophy-winning SB series of seaplanes and the eponymous Spitfire.

The other companies that later formed BAC could not trace their lineage back quite as far as Vickers, but all had contributed greatly to aviation in England since their inception. Bristol Aviation Company began in 1910 and over the next 50 years produced several aircraft designs that became household names. More than 5,000 two-seat F.2B Bristol Fighters were produced between 1916 and 1927, enjoying success over France during World War One, and in the interwar years formed the backbone of RAF operations, especially in the Middle East, until retirement in 1932. The Bristol Bulldog was a key part of Fighter Command's squadrons during the 1930s, while the Beaufighter gained a reputation as a fearsome anti-shipping aircraft during World War Two and played a part in the development of radar-equipped night fighters. Bristol's engine division also contributed numerous piston engines for use in its own aircraft and those of other manufacturers over the years. Post-war, the company incorporated a helicopter division under the directorship of rotorcraft pioneer Raoul Hafner, before an eventual merger with Westland.

Hunting Percival only came into being in 1944, when Hunting Engineering acquired Percival Aircraft, but Percival itself dated back to the early 1930s and had designed a range of single and twin-engine monoplanes for both the civil and later military markets. Some of those for the latter became RAF trainers and light transports, the twin-engine Pembroke and Sea Prince serving the RAF and Royal Navy into the 1970s and '80s.

English Electric (EE) was formed in 1918 from five companies involved in manufacturing armaments during World War One. In terms of designing and manufacturing its own aircraft, however, little remains of what was produced after that year, other than a single Wren ultralight of the early 1920s. Even so, the company had gathered experience building aircraft under licence, especially the wartime production of Hampden and Halifax bombers for Handley Page, both from a newly built site at Samlesbury near Blackburn. Post-war, EE purchased the former RAF base at Warton near Preston and, under designers W.E.W. 'Teddy' Petter and Freddie Page, designed and built two iconic aircraft of the 'Golden Age', the Canberra and the Lightning. From EE, through BAC and finally British Aerospace, Warton became and still is the home for Britain's advanced military aircraft projects.

Saunders-Roe (Saro) was well known as a builder of flying boats, but was another company with limited production volumes, even of its own designs; the 31 Saro Londons built for the RAF in the mid 1930s were the most numerous, but this company was another to provide volume production

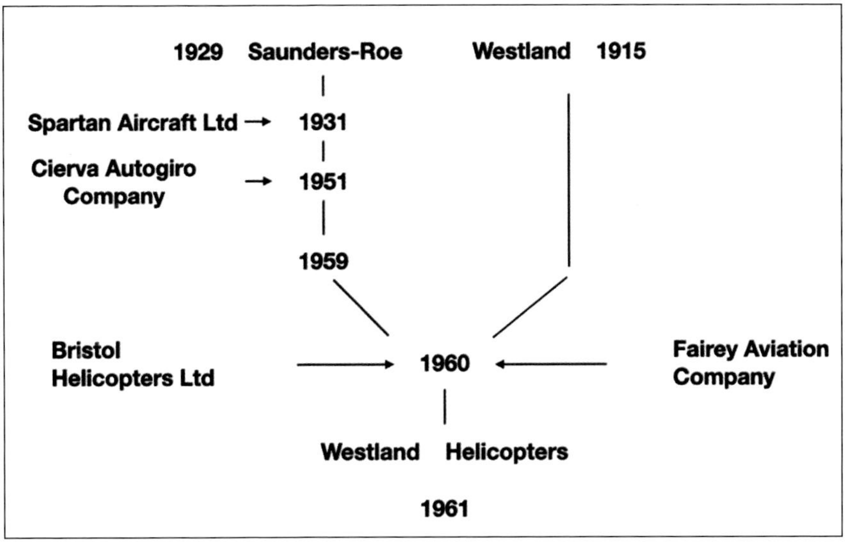

Westland's family tree, unifying Britain's helicopter-building capabilities under a single entity.

for others, in this case, the Supermarine Walrus and Sea Otter amphibians. With the age of flying boats virtually over, and despite producing the huge turboprop-powered Princess and considerably smaller SR.A/1 seaplane jet fighter, Saro diversified after the purchase of the Cierva Autogyro Company and began helicopter production. It branched out further to become Britain's early builder of hovercraft, while at the same time, winning a government contract to build a mixed-power development aircraft.

In contrast, since its inception in 1915, Fairey had built hundreds of aircraft and, much like Grumman became in the USA, was predominantly known for the aircraft supplied to the Royal Navy. During the inter-war period, Fairey produced various marks of the III Series whose service spanned from 1917 into the early 1940s, while the famous Swordfish, Albacore, Fulmar, Barracuda and finally the Firefly all served from the aircraft carriers of the Senior Service. The same could be said for Westland, which began as many did during World War One, producing aircraft to meet government-issued specifications, but this company then went on to develop and build its own aircraft designs. The Wapiti served the RAF well during the 1920s and '30s, especially in the Middle East, while the Lysander gained fame as a special operations 'taxi', depositing agents into France during World War Two, and the original but short-lived Whirlwind became the first all-cannon-armed RAF aircraft.

In 1960, Westland, along with the helicopter division of Bristol Aircraft and all Fairey aviation, merged to became Westland Helicopters. Prior to this,

Westland had signed what would become a hugely lucrative and successful licensing agreement with Sikorsky, allowing it to build and then adapt Sikorsky's helicopter designs for use with British engines and systems. Westland went on to produce the majority of rotorcraft for the British armed forces for more than 20 years[1]; all those following the Scout/Wasp were licence-built Sikorsky designs until the Anglo-French Lynx, Puma and Gazelle of the 1970s.

It's likely that all the individual companies would have carried on in their original form for some years, until mergers or takeovers came along anyway, as a result of commercial necessity or an offer they couldn't refuse. However, it is doubtful that many in their respective boardrooms anticipated that the offer they couldn't refuse would be coming from the British government, essentially obliging them to merge with other companies to tender for government contracts, and in this way at least have a chance of stay in the business of building aircraft; the other alternative was that the government won't be offering these companies any further contracts at all, in which case they would be on their own.

Those keen of eye will have realised that the one company missing from those mentioned above was the one that refused to merge, Handley Page. As a result, the company was excluded from direct government contract offers, which led to its decline. Despite the success of the Victor V-bomber, the last government contract awarded to Handley Page, albeit indirectly,[2] was for the single HP115 research aircraft. This was despite the preference of the RAF for the HP Herald as a tactical transport, a requirement fulfilled eventually by the Andover.

Frederick Handley Page founded the company bearing his name in 1909, building up a company that became synonymous with the O/100 and O/400 bombers of World War One. These aircraft evolved into the classic HP.42 passenger biplanes operated by Imperial Airways during the inter-war period, while the Halifax bomber of World War Two, which shouldered the burden of the night-time offensive against the Reich alongside the Lancaster and Stirling. What is usually forgotten, however, is that the last aircraft to initially bear the Handley Page name was the small twin-turboprop Jetstream. Exclusion from bidding for government orders saw Handley Page eventually enter liquidation in March 1970, with production of the Jetstream taken over by Scottish Aviation (SA). Much like Short Brothers in Belfast, SA was a regional company that the government exempted from mandated mergers. Rather ironically, once SA took over the Jetstream line, the government saw fit to buy

1. The only non-Westland type was the Aerospatiale Alouette II, operated by the Army from 1961 until retirement in 1988.

2. The HP115 was subcontracted to Handley Page by BAC.

these aircraft for the RAF in 1972, with a later additional batch for the Royal Navy. The Jetstream enjoyed further success in the civilian market when built under the British Aerospace banner.

Where once there were many manufacturers, by 1960 there were three major British aviation conglomerates; Hawker Siddeley, BAC (both of which had guided-weapon divisions in addition to aircraft), and Westland Helicopters, plus Shorts and Scottish Aviation, all bidding for government or airline contracts.

Whether the whole rationalisation and amalgamation process was a success or not is hard to tell in retrospect. In terms of number of aircraft produced by the three entities, Hawker Siddeley could be considered to have achieved the most, with Hunter production approaching 2,000 on its own, alongside the Buccaneer, Harrier, Nimrod, Andover and finally, the Hawk, prior to the final merger into another mandated union. This time all the non-rotorcraft companies were incorporated into British Aerospace.

Westland, having a monopoly over helicopter production, was always going to succeed, though still ran into financial troubles in the 1980s, whereas BAC, once production of Canberra and Lightning had finished, suffered mixed results. With more civil than military aircraft as part of its catalogue, BAC was at the mercy of Britain's airlines, which were just as likely to change their minds as the armed forces were forever doing, even after BAC had already gone the extra mile to give BEA or BOAC management exactly what they said

The first of a very successful line of Sikorsky designs built under licence by Westland and 'anglicised' by replacing the American engines with a British equivalent, was the WS-51 Dragonfly. HR.5 WG719 is at the Helicopter Museum.

they wanted in the first place. More significantly, building to suit the whims of the national airlines repeatedly left the company with designs generally unsuited to world markets without expensive alterations, handicapping them in volume and hence cost when set against the products of Boeing or Douglas. This situation was repeatedly made worse when designs that were originally offered were more akin to those needed worldwide and belatedly wanted by BOAC and BEA far too late in the day.

Beyond the period covered in this book, one lesson to emerge from the demise of TSR.2 does appear to have been learnt; if the aircraft in question was developed and built in tandem with another country, it was much harder for the government to cancel the project! Consequently, Concorde, Jaguar and Tornado were seen through to completion, when at times it looked like the government of the day was having second thoughts.

Nearly 60 years after the end of the 'Golden Age' in British aircraft manufacturing, and now under the auspices of BAe, there was contemplation of a purely British-designed and built aircraft in the shape of Project Tempest, but such is the cost and complexity of modern aircraft that this too has become multinational. Italy, the previous partner on Tornado and Typhoon, is involved again, plus Japan for the first time; the possibility exists that other countries may join the endeavour before a production version arrives. If the proposed schedule is maintained, the prototype is pencilled in for first flight in 2035. However, at the time of writing, no official guarantees exist that the project will be seen to completion; one wonders if history will once again be repeated.

Chapter 2

Bristol Type 188

Just ten years after the end of World War Two, plans emerged for a high-altitude, high-speed reconnaissance aircraft, potentially capable of Mach 3 with Mach 2.5 cruising speed. This represented speculation by the officials in Whitehall as to how far the aircraft and engine industries could advance in the immediate future, or at least how much the upper ranks of the RAF believed what the representatives from these same companies were telling them. Both outcomes were contingent, of course, on how much money could be got out of the government in the first place.

Operational Requirement 330 (OR.330) imagined just such an aircraft, whose primary mission as initially envisaged would have been radar mapping of targets for the V-Force, especially once these aircraft possessed stand-off weapons. Requests for proposals were issued to industry in 1954, with designs returned from Handley Page, Vickers and Avro. The order to create this dream machine was subsequently awarded to Avro for its Type 730. This

A Bristol 188 in the centre of the test flight hall at the RAF Museum Midlands. The RAF Museum's disposal plans may see this aircraft loaned to Aerospace Bristol.

was a futuristic-looking tail-less delta-winged arrow with front canards and power coming from eight Armstrong Siddeley P.176 turbo jets mounted as two sets of four, grouped within a pod in each wing. It was expected that Avro would utilise the same honeycombed high-tensile steel construction it had developed for the 720 rocket/jet fighter, providing strength whilst keeping airframe weight down. This was crucial, as OR.330 specified a range of 4,000–4,500 miles without in-flight refuelling, yet the aircraft needed engines powerful enough to push it to a sustained Mach 2.5, resulting in an undoubted thirst. Consequently, half the take-off weight would have been jet fuel, but leaving sufficient space and payload capacity to house the lengthy radar array, assigned the rainbow code 'Red Drover', that was needed to carry out the intended mission. To make things worse, even before metal had been cut on any sort of prototype, the RAF had decided to add strategic bombing to the already extensive requirements, seeing the 730 as a potential successor to the V-bombers. Changes to this extent could easily have led to a complete redesign of the fuselage to accommodate a weapons bay as a minimum, but the RAF chiefs still expressed disappointment that the projected in-service date was 1964.

Supersonic flight is not simply a matter of mating a sleek low-drag airframe with powerful engines, though at its minimum it is precisely that. The friction or kinetic heating caused by air rushing past the thin metal skin of an aircraft at just over 2,000mph results in the airframe heating if this speed is sustained. At Mach 2.0, this temperature can reach 100°C, but at Mach 2.5, it nearly doubles to 190°C. With few aircraft able to achieve anything close to these speeds at the time (the Mach 2.0-capable EE Lightning wouldn't enter service for a further six years), more information was needed on how these temperatures affected the metal alloy structures used to date. The test vehicle to explore these effects was the Bristol Type 188, and here begins a tale of woe, because the Type 188 simply never got close to meeting the goals set for it.

To cope with the expected aircraft skin temperatures, Bristol chose a specific type of chromium-hardened stainless steel and opted to use a new-at-the-time argon gas 'puddle welding' technique to marry some of the sections of the airframe together. This caused long delays until the process produced welds of sufficient quality. Similarly, despite the aircraft originally being designed to use Rolls-Royce Avon RA24R engines (later fitted in the RAF's Lightnings), those eventually fitted were descended from one of the great hopes of the UK jet engine industry, the de Havilland Gyron, in this case a pair of reheated Gyron Junior turbo jets, producing 14,000lb thrust with reheat engaged, 10,000lb without. The un-reheated Gyron Junior flew in the Mk 1 Blackburn Buccaneer, while the reheated version was intended, along with a Spectre rocket motor, to power the Saunders-Roe SR.177. Progress with the reheated version was slow, with impetus lost and more delays incurred following the cancellation of the SR.177. The choice of engine also necessitated a complete redesign of the nacelles to house them compared to the Avon; another time-consuming exercise.

After eight years of setbacks and £20m in funding, the first flight of the Bristol 188 finally took place on 14 April 1962, when XF923 took off from Bristol's runway at Filton, with Chief Test Pilot Godfrey L. Auty at the controls. By this time, the Avro 730, whose development it was supposed to advance, had long since been cancelled in the 1957 Defence White Paper. With government financial constraints and despite the need to better understand kinetic heating, it is hard to understand why construction of the Bristol 188 continued.

Despite easily passing the sound barrier, the best speed that the 188 could manage was a hugely disappointing, yet rather ironic Mach 1.88 at 36,000ft, a far cry from the Mach 2.5 it was supposed to achieve. Problems with the Gyron Junior engines made controlling the machine at these speeds virtually impossible, while their prodigious fuel consumption precluded extended flight times at high Mach numbers; a mere two minutes was possible from a total endurance of around 25 minutes. Consequently, the short high-speed flight duration failed to give adequate time for kinetic heating to have any impact on the airframe, making the suite of paper and magnetic tape recorders mounted in the fuselage to capture all the data somewhat redundant.

A view of the Avro 730 in model form at the RAF Museum Midlands.

Bristol 188 being checked over. Note that it seems to be missing the forward cowling off the engine nacelle.

Bristol 188 under construction at Filton.

***Above and below*: Two views of the surviving Type 188, XF926, resplendent in the Test Flight Hall at the RAF Museum Midlands, Cosford. The other airframe built, XF923, was scrapped at Foulness in 1966.**

The solution would have been to re-engine the aircraft again, possibly using the Avon it should have had in the first place, but by then it was all far too late, even though a second aircraft, XF926, had been completed and flown. Flying ceased in January 1964, after which both aircraft were taken to Shoeburyness in Suffolk to serve as gunnery targets.

XF926 was rescued from this ungallant end and originally sat resplendent in the middle of the Test Flight display hall at the RAF Museum, Cosford, before the museum began a major reorganisation in 2025; a polished stainless-steel monument to how badly things can go wrong when you're trying to push the boundaries of flight.

As for high-altitude, high-speed reconnaissance, just ten months after the Type 188 was struck off charge, Lockheed test pilot Bob Gilliland flew the first of an eventual 32 Lockheed SR-71 Blackbirds, the USAF's hugely impressive, considerably more successful and still officially the fastest air-breathing jet-powered aircraft in the world. In 1976, one set a new world speed record at 2,193.2mph.

Chapter 3

Hunting H.126

In order to test ideas emanating from the National Gas Turbine Establishment (NGTE), part of the Royal Aircraft Establishment, for 'jet flaps' or 'blown flaps', Hunting was awarded a construction contract in 1959, under Ministry of Aviation specification ER.189D, for two demonstrators featuring this technology and created a unique aircraft.

The H.126 is powered by a single Bristol Siddeley Orpheus engine of 4,000lb thrust; nothing unusual so far. However, a series of ducts takes the exhaust from the engine and pipes it to a distribution manifold, from which 40 per cent of the gases are directed across 16 nozzles spanning virtually the entire trailing edges of both wings. Another set of ducts takes a further 10 per cent to the control nozzles on the wingtips and in the tail; a similar system to that of the Rolls-Royce TMR and Short SC.1. Although this leaves 50 per cent of the theoretical engine output available for forward propulsion, in reality it doesn't provide a great deal of impetus, so while top speed was never going to make any headlines, the jet flaps resulted in take-off being achieved at around 32mph, a figure that even light propeller-driven aircraft would find hard to beat.

Although two airframes were requested initially, the second, XN719, which would have been built with a different amount of wing dihedral, was never completed. This left the first aircraft, XN714, to make a maiden flight in March 1963. It went on to carry out 100 test flights, including a demonstration flight at the Paris airshow in 1966 after slowly making its way there under its own power; ultimately, it accrued just over 140 flying hours in total and continued until November 1967, when it undertook its last flight.

During 1968, XN714 was sent by road to Hawker's factory at Brough and by April 1969 it had been dismantled for transport by an RAF Short Belfast to the NASA Ames Flight Test Centre at Moffett Field, California, USA, where a series of tests was carried out using NASA's large wind tunnel facility. Details of what these tests were for, or their outcome, do not appear to have been made public. The aircraft returned to the UK in May 1970, where it spent the next three and a half years in transport crates, eventually being struck off charge in June 1972. Finally, it was taken by road in April 1974 to the RAF Museum at Cosford, where it currently remains.

With the complex arrangements for the engine exhausts, along with the ducting for the hot gases requiring heat shielding, there were distinct weight penalties incurred for any potential aircraft using this system. Unsurprisingly, and despite the undoubted efficiency of the jet flaps at low speeds, it was never adopted for any other aircraft.

***Above and below*: Two views of the Hunting H.126 at the RAF Museum Midlands.**

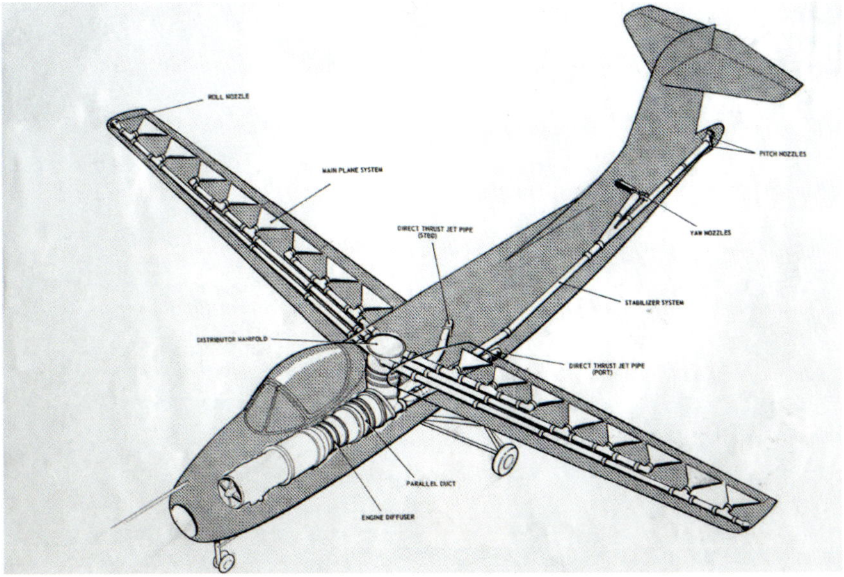

A diagram showing how the jet efflux is routed to the wings and control ducts.

The blown flaps of the H.126 stretch virtually the entire length of the wings. What appears to be a wing brace is the exhaust feed from the distribution manifold to the outer flap units.

Looking rather ungainly next to the SR.53 and the BAe EAP, the Hunting 126 nonetheless performed exactly as designed.

The H.126 in flight.

Possibly April 1969, when the H.126 was packed aboard an RAF Short Belfast transport for delivery to NASA for wind tunnel tests.

Chapter 4

Training Syllabus

Hunting Percival P.84 Jet Provost

Away from the trailblazing, technologically frontier-pushing projects of the time, the job of training pilots for the RAF carried on as normal, but some foresaw the possibility of an all-jet-powered training syllabus and a suitable aircraft to fill that position looming on the horizon. Hunting Percival had already supplied the RAF with the piston-engine Provost and began developing a jet-powered version as a private venture, giving this company a head start over potential competitors.

Good contacts within the RAF training command provided Hunting with details on how senior staff envisaged the direction of the flying training programme. This helped evolve what became a very successful and long-lived aircraft. The Jet Provost was built with an intention from the outset to replace propeller-driven trainers from day one, as opposed to being something a pilot would move on to once they'd flown a propeller aircraft. In this way, the likelihood of nurturing propeller-relevant 'habits' would be removed. The aircraft was also flexible enough to permit its use in all aspects of flying training. As a result, one of its best characteristics was that it was easy to fly, while the longevity derived from that was equally as easy to maintain.

A pair of Jet Provost T3 from 2 Flying Training School at RAF Syerston in 1960.

Government funding towards the development eventually appeared in 1953 and on 16 June 1954, P.84 Jet Provost prototype XD674 took to the skies for the first time above the Hunting factory at Luton airport, piloted by Dick Wheldon. Its ancestry from the piston Provost was apparent, owing to the adoption of the same tail planes and wings, and even re-using the main landing gear, though switched to a tricycle arrangement. Power came from an Armstrong Siddeley Viper turbojet fed from small intakes either side of the cockpit, which was in the side-by-side arrangement preferred at that time. After XD674 passed Hunting's own testing process, it was sent for RAF trials at Boscombe Down, where the first changes requested were to shorten the undercarriage to provide a smoother ride, especially over semi-prepared surfaces.

The RAF was sufficiently impressed by the Jet Provost that a pre-production batch of ten aircraft was commissioned, with the first three assigned to the Central Flying School. Here several student pilots used the aircraft as part of their training syllabus, and such was the improvement in the rate of learning that the RAF requested a production version. The Jet Provost T3 duly entered RAF service following a first flight in June 1958, incorporating additional changes from the pre-production batch alongside the previously requested shortened undercarriage. These included a more powerful Viper engine and the addition of a dorsal fillet, with further, more subtle structural and systems alterations.

Over the next four years, 201 aircraft were delivered to the RAF, by which time a T4 version had already flown; again, this upgrade was primarily for the incorporation of an uprated Viper engine. In total, 198 T4s were built.

Hunting subsequently became part of BAC, and under its auspices, XS230, the penultimate T4 off the production line, was converted into the T5 prototype and flown in 1967. Once testing was complete, XS230 was destined to serve with the Empire Test Pilots School at Boscombe Down. Today, it is still on the UK civil aircraft register as G-VIVM and resides at North Weald airfield. In addition to the now customary engine power increase, the T5 included cabin pressurisation for the first time, and a new, more distinctive sloped nose cone that also housed the landing lights. BAC later developed the T5 into the Strikemaster trainer and light attack aircraft for the export market, equipped with tip-tanks, wing hardpoints and mountings for a pair of machine guns under the jet intakes. Modest sales followed, with 146 sold worldwide to customers including Oman and Ecuador; some saw combat and casualties.

The handling qualities of the Jet Provost, usually referred to simply as the JP, also facilitated its adoption by numerous RAF display teams during its career. This went as far back as 1958, when a four-ship team from the Central Flying School (CFS) at Little Rissington called 'The Sparrows' flew

The P.84 prototype Jet Provost XD674, now housed at the RAF Museum Midlands, Cosford.

Note the original rather tall undercarriage legs on the P.84, a result of the main wheel legs being borrowed from the earlier Piston Provost, which is hiding behind in this shot.

Jet Provost T1 XD694. This was the last one of ten built for the RAF from a total of twelve constructed, Percival retaining the final two. Compare this version to the later marks; the undercarriage has been shortened from the prototype, but lacks the enlarged fin and dorsal fillet introduced after.

Jet Provost T2 in natural metal finish and `B' Class identity G-23-1, as it was owned by Percival Aircraft. It later became XN117 and was converted to T3 specification, before being scrapped in the 1960s.

Norfolk and Suffolk Aviation Museum's Jet Provost T3A, XN500, in the unusual colour scheme of the Oxford Air Training School.

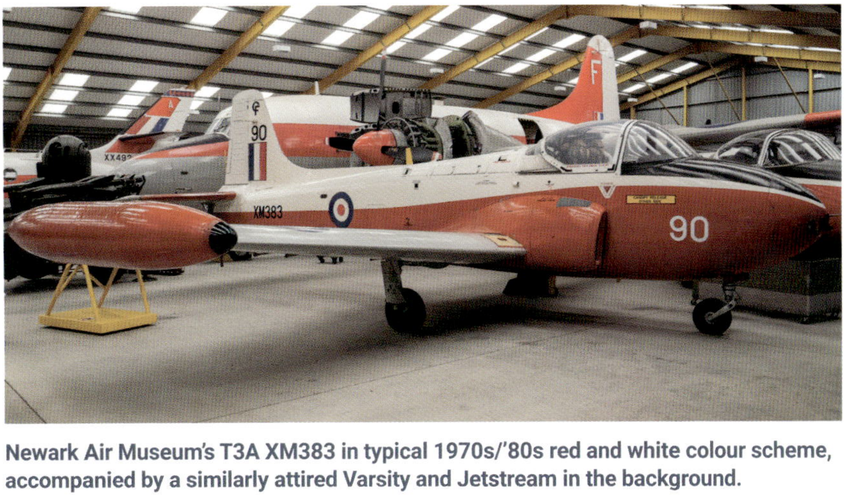

Newark Air Museum's T3A XM383 in typical 1970s/'80s red and white colour scheme, accompanied by a similarly attired Varsity and Jetstream in the background.

for a season, through the 1960s, when another four ship CFS team, 'The Red Pelicans', performed for two seasons in suitably coloured T4s. Even after the Red Arrows became the official RAF display team, JP teams continued to appear on the airshow circuit during the 1970s, with the likes of the 'Gemini Pair', 'The Macaws' and 'The Poachers' (all seen by the author on multiple occasions) wooing the crowds.

The RAF retired the Jet Provost in 1993, with many going to museums for display, but others have passed into private hands and have been kept airworthy, a tribute to being relatively cheap to run in jet aircraft terms, but also to their simplicity of maintenance and ease of access for servicing. Multiple airframes found their way to the RAF training establishment for the same reason.

Folland Gnat

Not content with being involved with designing both the Canberra and the Lightning, W.E.W. 'Teddy' Petter went in a totally different direction with what became his final design. Following his departure from English Electric after long-term disagreements with general manager Arthur Sheffield, Petter was hired as chief designer by Henry Folland, and succeeded him as managing director of Folland Aircraft less than a year later. Though he successfully poached a few of his former staff from English Electric to create his new design team, the first government contract that came its way became something of a poison chalice, the *Red Dean* air-to-air missile (see later). This eventually

Spot the difference. Folland's Gnat Prototype G-39-2 is on the left and Midge G-39-1 is on the right. The Gnat was marginally longer to accommodate the more powerful Orpheus engine and had a slightly taller nose undercarriage leg to raise the wing incidence on take-off.

wound up in the hands of Vickers after Folland terminated the contract in November 1951, unable to build anything close to meeting the requirements.

Influenced by reports of the Soviet Air Force's fleet of Tu-4 bombers (clones of the Boeing B-29, a number of which had crashed or landed in Russia during the war and Stalin had refused to hand back, so allowing Tupolev to reverse-engineer them), Petter went on something of a crusade to design and promote light fighters. Where the Hawker Hunter weighed in at just over 21,000lb and Petter's own Lightning nearly double that when fully loaded, the idea that he could produce an effective combat aircraft closer to 5,000lb underlines what he meant by a light fighter. Seeing this as a more affordable, but still a capable aircraft, Petter suggested that for the cost of one normal-sized fighter, you could have five light fighters. Similar thinking spawned the Northrop F-5 Freedom Fighter, which also started as a private venture.

Unable to find an ideal power plant, with high thrust but low weight, Petter opted to use the Armstrong Siddeley Viper in what became the Fo-139 Midge, the sole example of which flew on 11 August 1954. As a proof of concept, with no official requirement for anything along the lines he was considering, this aircraft's development was wholly company funded. It was considered a success; it handled well, went transonic in a dive and garnered praise from most who flew it. Unfortunately, on 26 September 1955, the sole Midge built was lost in a crash along with the Swiss pilot evaluating it.

Shortly before this, and after talks with Ralph Hooper at Bristol Siddeley, Petter finally acquired a more powerful engine, the Orpheus. This became the missing piece to complete the Fo-140 Gnat, which had been developed alongside the Midge. The Gnat incorporated a slightly increased wing area in comparison to the Midge, plus larger air intakes to provide sufficient air flow for the Orpheus, which simultaneously allowed space to house a 30mm Aden cannon in each engine intake lip.

The prototype, G-39-2 (civil registered, as opposed to bearing an RAF/RN Ministry-allocated serial number, as it was not government funded), first flew in September 1955 and again was a private-venture aircraft, though a further six were later ordered by the Ministry of Supply for evaluation and were publicly funded. Although no orders initially came from the UK government, the Gnat garnered interest and customers elsewhere, with orders from Finland for 13 aircraft, as well as an evaluation pair for Yugoslavia[3]. More significantly, a great deal of interest came from India with an order for 25 aircraft direct

3. After one of the Yugoslav aircraft crashed in 1958, this country showed no further interest. The remaining aircraft is on display at a museum in Serbia.

from Folland, plus 15 in kit form along with licensing for production in India by Hindustan Aeronautics Limited (HAL). Around 200 were eventually built.

At first, both India and Finland found the Gnat to be troublesome, primarily suffering hydraulic problems. The Finns' experience was mainly due to the system being unable to cope with the cold climate, whilst both also criticised it for somewhat idiosyncratic handling at certain points in the flight envelope, especially when raising or deploying the undercarriage. The doors for this acted as an airbrake and the resultant airflow changes caused problems, especially for new pilots, leading to a number of crashes. Both air arms also reported that the Gnat required significant maintenance hours to keep it flying. Still, once the teething troubles were ironed out, both air forces used the Gnat extensively. India made full use of the agility and versatility it offered, taking the Gnat to war twice against Pakistan in 1965 and 1971, gaining air-to-air kills against Pakistani Air Force Sabres, while also carrying out ground-attack, photo reconnaissance and anti-shipping missions with the diminutive jet. HAL later developed the Gnat into the slightly larger but more capable 'Ajeet', though increased weight cost the type some of the original's agility.

The Gnat did finally find success in its home market, though probably later than Folland hoped, leading the company to conclude that this was a not-so-subtle hint to merge with one of the other manufacturers following the 1957 White Paper. This duly took place, and Folland became part of Hawker-Siddeley in 1959. At the same time, Petter realised that although he got on well with Hawker's chief designer Sir Sydney Camm, the clash of personalities was unlikely to work on a day-to-day basis. Consequently, he resigned the same year and simply walked away from the aircraft industry.

The Gnat, of course, did not serve with the RAF in the light fighter role for which it was originally conceived, but as a trainer, built around specification T.185D. This provided a transitional aircraft for pilots going from the rather tame Vampire T11 and later the Jet Provost to the vastly more powerful Lightning. The changes made to meet the RAF's requirements also alleviated the issues with flying characteristic problems reported by other operators, mostly due to the adoption of more conventional wing control surfaces and the undercarriage doors no longer acting as air brakes. Marginally larger wings allowed additional fuel tankage to replace that lost by the addition of the second seat, but adoption of permanent slipper tanks at the mid-wing position further countered this to some extent. One complaint from all operators of the Gnat, however, came from both ground crews and pilots; there was not much room to work. For the pilots, some found the cockpit too small altogether and they had to be assigned to different aircraft, whilst instructors found the rear cockpit cramped. Where ground crews

***Above*: Close to home: Gnat XM693 mounted outside Hamble Aerostructures sports facility, not far from the original Folland factory. This was the third T.1 prototype.**

***Right*: Finnish Air Force Gnat F.1s on a snowy dispersal. The diminutive size of Folland's lightweight fighter can be gauged by the ground crew stood next to the bottom aircraft.**

were concerned, access to systems was awkward and trying to get different trades to access components in the same general area was doubly difficult, as many trainees discovered when retired Gnats found their way to RAF training schools; RAF Halton had several instructional Gnats well into the 1980s. Such were some of the aircraft's idiosyncrasies; it was also said by pilots that as opposed to learning to fly in general, you very much learnt to fly the Gnat.

The 105 Gnats built for the RAF primarily operated from 4 Flight Training School (FTS) at RAF Valley, to train pilots assigned to fast jets for around a further 70 hours, after their initial qualifications were gained on the Jet Provost. They then moved to Operational Conversion Units for the aircraft types to which they'd been posted.

From its initial introduction in 1962, until replaced by the Hawk, the Gnat was a familiar sight at air shows, but fame truly arrived with the formation of two noted aerobatic teams. First came the Yellowjacks in 1964, in suitably bright yellow-liveried Gnats, but these were overshadowed by the Red Arrows, the RAF's officially funded display team, which formed later the same year. The Gnat was flown all the way up to 1980, but was finally relinquished and replaced with the Hawk T.1.

Gnat F.1 XK724, the first of six evaluation aircraft ordered by the Ministry of Supply. Note that the wing is removed as a single unit, as well as the gun port with vents on the air intake for one of the pair of 30mm Aden cannons fitted. This aircraft is now housed at the Midland Air Museum.

Chapter 5

Short SB.4

One benefit of companies not having the luxury of computer modelling during the 1950s was the number of unique aircraft produced, simply because building something and trying to fly it was the only real way to find out if it worked. Wind tunnels, scale models and calculations could only go so far.

The Short SB.4 and earlier SB.1 glider could definitely be described as being unique; both were built to study the properties of the aero-isoclinic wing proposed by Professor G.T.R. Hill. During the 1920s, Professor Hill had designed a series of rather strange-looking tail-less aircraft for Westland, including the Hill Pterodactyl 1A which now hangs in the Science Museum in London. His aero-isoclinic wing was designed to study the effects of maintaining a constant angle of incidence despite wing-flexing loads on tail-less aircraft, along with other problems associated with swept wings, including flutter and tip stall.

Designed by Short's chief designer David Keith-Lucas, the SB.1 glider first flew via a winch launch on 14 July 1951, with the company's chief test pilot Tom Brooke-Smith at the controls. He managed a second flight via a towed launch the following day. However, disaster struck on the third launch, after buffeting from turbulence caused by the tow plane forced Brooke-Smith to cast off at low altitude. Despite his best efforts, the glider suffered a heavy nose-first landing, seriously injuring Brooke-Smith and severely damaging the glider.

While Brooke-Smith recuperated, the decision was made to rebuild the SB.1, but not as a glider; it became the SB.4 Sherpa, with a pair of Blackburn Turbomeca Palas turbojets providing a modest 350lb thrust each. As this was a tail-less design, elevator surfaces along with ailerons were in the wings, but control was achieved via a very unusual method. Rather than individual moving surfaces as found in a conventional aircraft, either split between wing and tail, or in other tail-less designs separate sections of the wing trailing edge, on the SB.4 the entire wingtip sections rotated together as elevators, or in opposition as ailerons.

Tom Brooke-Smith recovered in time to pilot the SB.4 once it was completed, undertaking the maiden flight on 4 October 1953, with continued evaluation

The Short SB.1 glider is launched on one of only three flights before crashing.

The SB.4 in flight, showing the dorsal air intake behind the cockpit and illustrating the unusual wing shape; just visible are the wing tips working together as elevators.

A close up of the SB.4 wing tips, with what appears to be a trim mechanism.

***Above*: The Short SB.4 now at the Ulster Aviation Collection. (Photograph courtesy Alan Jarden)**

***Right*: New wing under construction; note the diminutive exhaust for the Turbomeca Palas jet engine. (Photo courtesy Mark J. Cairns)**

flights through to mid-1954 and with no further accidents. Although the SB.4 flew well enough, the wing design didn't live up to expectation, with no performance benefit over a conventional wing design, and it was never adopted elsewhere, though Short at one point proposed an aero-isoclinic winged Supermarine Swift. Its P.13 design, submitted for specification NA.39 and leading to the Buccaneer, also used a similar wing, but neither progressed off the drawing board.

The single SB.4 spent the remainder of its life moving between several advanced technology institutes as an instructional airframe, until being finally grounded in 1964. It has since been completely renovated by the Ulster Aviation Society in Northern Ireland and is on display there.

Chapter 6

Saunders-Roe SR.A/1

Some of the aircraft still on drawing boards of the late 1940s were originally conceived with a war still raging, and the expectation of a difficult fight in the Far East to defeat the Japanese. Fortunately, or unfortunately, depending on how you view it, the dropping of atomic bombs on Hiroshima and Nagasaki brought hostilities to a more abrupt halt than expected. Consequently, some designs for new models, and certainly the orders for existing aircraft were simply cancelled or dramatically cut. However, one design ultimately to make it out of the drawing office and into the air was the Saunders-Roe SR.A/1, which became the world's first jet-powered flying-boat fighter.

The original request for a flying-boat fighter-bomber came from the Admiralty, for an aircraft able to operate from sheltered coastal waters. The

SR.A/1 overflies the Isle of Wight coast; all three aircraft were built at Saunders-Roe's factory at Cowes.

thinking was perhaps along the lines that, given the huge expanse of the Pacific Ocean in particular, and the time needed to either restore or build an airstrip in captured territory for operational use by land-based combat aircraft, the ability to turn any quiet atoll or calm-watered inlet into a fighter base in comparatively no time seemed like a reasonable idea. Blackburn Aircraft had successfully tendered to meet the requirement in 1942, its proposal designated N2/42, but its use of the heavy Napier Sabre IV piston engine and a fuselage that incorporated a hull that could be raised and lowered using a scissor mechanism, led the Maritime Aircraft Experimental Establishment (MAEE) to conclude that it would struggle to actually get into the air.

Saro proposed its design to the Air Ministry in 1943 as the SR.44, and this was finally approved after a number of ministry-suggested design changes, resulting in specification E.6/44 with an order for three prototypes. The end of the war and Saunders-Roe's development work on the huge Princess flying boat led to a less-than-speedy appearance of the SR.A/1, but it finally flew on 16 July 1947 at the hands of test pilot Geoffrey Tyson. Other demonstrations followed, including a display at the SBAC Show at Farnborough in 1948, where the machine demonstrated agility and performance considerably better than hinted at by its rather ungainly appearance.

Power for the SR.A/1, unofficially dubbed the 'Squirt', was provided by a pair of Metropolitan-Vickers (Metro-Vick) Beryl F.2/4 axial-flow turbojets, which for

The only remaining SR.A/1 with P (Prototype) series markings. The engine on the yellow dolly in front is one of the Metropolitan-Vickers Beryl FB.2s fitted to the SR.A/1, showing its relatively compact size. (Author's SR.A/1 photos all courtesy Solent Sky, Southampton)

the time was quite an advanced design, eventually providing 4,000lb of thrust compared to the 2,400lb of the contemporary Rolls-Royce Derwent found in the Gloster Meteor. If production aircraft had been ordered, however, they would have needed a different power plant, as Metropolitan-Vickers decided to pull out of the jet engine business, its designs being acquired by Armstrong-Siddeley. Thankfully, the work carried out developing the F.2 didn't go to waste, as it provided the core for Armstrong Siddeley's very successful Sapphire series of engines.

Three prototype SR.A/1s were duly built, although two crashed during the four years of development flying that were undertaken. In the first accident, the aircraft was being piloted by the highly experienced test pilot Lt Cmdr Eric 'Winkle' Brown, and it nearly cost him his life! Flying from Saro's facilities at Cowes on the Isle of Wight, Brown decided to land near the mouth of the River Medina, but a submerged object in the path of his

The jet pipes extend off the high-mounted wing and well out of the way of any water. The cockpit, seemingly perched amidships, would have provided good all-round views when fitted with the fully glazed canopy, which would have replaced the one fitted here.

The SR.A/1 is much larger than you expect it to be, the boat hull adding considerable height, while accessing the cockpit must provide something of a challenge wearing flight gear on land, let alone afloat.

Air-to-air photo call, with TG263 above the clouds. Note the wing float in the retracted position and the later clear canopy.

landing punctured the hull of the aircraft and sheared off the starboard float. Only the prompt actions of Geoffrey Tyson, who was aboard Saro's company launch, saved the day; he rescued the unconscious Brown from the aircraft, moments before it sank. Quite fittingly, therefore, the SR.A/1 was also the first aircraft to have production Martin Baker ejector seats fitted, though 'Zero-Zero' seats, (zero speed and zero altitude), were some years off.

After the SR.A/1 was placed into storage in 1950, interest in the aquatic fighter was briefly reignited by the outbreak of the Korean War, with Saro hoping to sell a developed version of the SR.A/1 for service in this theatre, but these came to naught, despite a prod to government by landing TG263 on the Thames and mooring it near the Southbank Festival site in June 1951.

The remaining SR.A/1 is an impressive centrepiece at the Solent Sky Museum in Southampton.

Chapter 7

Saunders-Roe SR.53

Mixed-power aircraft, where a jet engine was supplemented with a rocket motor, were for a time seen as the only way to achieve fast climb and high-altitude operations. The first British mixed-power trials aircraft was the Hawker P.1072, built from Hawker's original P.1040 Sea Hawk prototype VP401 and first flown in November 1950. This sported a liquid oxygen and water/methanol-fuelled Armstrong Siddeley Snarler rocket motor in a slightly modified tail unit, in addition to the usual Nene, with the space between the bifurcated jet exhausts providing adequate room for the installation. Although the rocket motor gave a 2,000lb instant increase in thrust, it came with a commensurate increase in fuel consumption. As for reliability, six flights were carried out before a minor explosion damaged the aircraft beyond repair. This hiccup didn't stop further funds and much effort being expended.

With a design department looking where to go next after its efforts with the Princess flying boat and SR.A/1, Saunders-Roe had initially suggested a transonic fighter flying boat, but found no takers, so turned its attention to a rocket-powered fighter able to reach 100,000ft. Further talks resulted in the issuing of a 1952 Ministry of Supply (MoS) specification F.124T, for a Mach

RAF Museum Midlands Short SR.53 with *Firestreak* missiles on the wingtips.

The prototype Hawker P.1040 Sea Hawk in later life, with the addition of an Armstrong Siddeley Snarler rocket motor in a modified tail assembly.

2-plus, fast-climbing point defence interceptor. The aircraft was required to attain 80,000ft and to be armed with a battery of folding fin rockets in a retractable housing, though this armament later gave way to a pair of heat-seeking *Firestreak* missiles once these became available. The official specification was later amended to F.138D following more changes and the aircraft was named the SR.53. The intention was to combat an expected threat from Soviet bombers, initially high subsonic and later supersonic, with the fighters deployed in suitable locations and scrambled to meet an impending threat once it had been detected by the newly implemented ROTOR radar system. Power for the SR.53 was provided by a High-Test Peroxide (HTP) and kerosene-fuelled de Havilland Spectre rocket motor producing 7,000lb of thrust and an Armstrong Siddeley Viper of a much more modest 1,640lb thrust, the pair sitting in a sleek fuselage with a T-tail and cropped delta wings.

With government policy at the time requiring competitive designs (widely acknowledged as a euphemism for having a back-up in case one failed to make the grade, but a huge drain on resources all the same) A.V. Roe was also contracted to provide a similar aircraft, the Avro 720. Meanwhile, the RAF was having doubts that it wanted a rocket-engined fighter in the first place, so was willing to wait for both the SR.53 and Avro 720 to fly before making any sort of final decision.

Ultimately, government cost-cutting and Avro's choice of liquid oxygen (LOX), methanol- and kerosene-fuelled Armstrong Siddeley Screamer rocket led to the Avro 720's cancellation, despite being potentially a year ahead of the SR.53 in terms of construction and expected first flight date. Having LOX tanks in a combat aircraft was now regarded, certainly by the Navy, as even more undesirable than tanks of HTP, especially within the confines of a carrier hangar, while the practicalities of having an aircraft needing three liquids for fuelling in an operational environment, including one with a distinctly unique set of handling requirements, would definitely have been something of a challenge.

It eventually took five years to get the first prototype SR.53, XD145, into the sky, on 16 May 1957, piloted by Sqn Ldr John Booth. Problems with the Spectre rocket motor, limitations on how many engineers could work within the confines of a small airframe and the very exacting limits on welding the tanks that housed highly volatile HTP rocket motor fuel all contributed to a longer than desired delivery period, to say nothing of meeting the scheduled delivery time. The lengthy gestation period did, however, give Saunders-Roe time to conduct extensive testing on the handling and storage of HTP. It was found that fuel tanks could be safely left filled for over a week, with few additional handling procedures needed, aside from highlighting the need for stringent cleanliness, as contact with only small amounts of organic matter causes spontaneous combustion. The company concluded that the fuel could easily be neutralised with water (water pistols were even included in special ground handling kits to cope with minor spillages!), suggesting that HTP posed no greater risk than normal jet fuel. Later photographs of RAF personnel handling the similarly fuelled Blue Steel missile wearing protective suits, with provision of emergency water baths and water bowsers on standby, would suggest that others failed to agree.

The SR.53 designed as a point defence interceptor, which restricted space for fuel and extensive avionics.

The two SR.53 prototypes completed a total of 42 flights between them and at least established that the jet-plus-rocket combination worked as advertised. XD145 was displayed at the SBAC Show at Farnborough in September 1957 to great effect. Unfortunately, after only 11 flights, XD151, the second prototype, crashed at Boscombe Down on 5 June 1958, killing Sqd Ldr Booth, when the aircraft failed to take off and collided with a concrete approach light. Despite the best efforts of both Saunders-Roe and the RAE Accident Investigation team, the subsequent board of enquiry never determined the actual cause of the crash, nor found evidence of pilot error. Following the crash, XD145 completed a further 14 flights, making 56 in total for both aircraft, with over 22 hours of flying time logged. However, after the last flight in October 1959, XD145 was grounded and never flew again. The aircraft had attained Mach 1.33, reached 55,000ft and achieved an impressive 29,000ft-per-minute rate of climb. Surprisingly, the performance did not quite meet what was expected, this being attributed to a 1,000lb thrust shortfall in the Spectre 1A rocket motor. Lt Cmdr Peter Lamb, who concluded the test flying, reported that the flying and handling qualities of the SR.53 itself were 'extremely docile and it was very pleasant to fly'. Saunders-Roe was officially informed that the whole programme had been terminated on 29 July 1960.

Displayed at the RAF Museum Midlands the first prototype XD145 sits resplendent in the classic overall gloss white finish, but the SR.53 story may have had a different ending.

During development of the SR.53, Saunders-Roe had been running a scheme in parallel to investigate further enhancements in mixed-power combat aircraft. The RAF's requirement (OR.301) and the MoS specification that prompted the building of SR.53 back in 1951, had envisaged a fully operational aircraft and not just a development vehicle, but on closer inspection, it was obvious that both sets of documents had a major flaw. Despite the specification being envisaged as a point defence interceptor, neither mandated fitment of air-to-air radar, leaving the pilot needing ground-based controllers to vector him and his aircraft into the general vicinity of a target, where he became reliant upon the Mk1 eyeball to acquire an enemy aircraft and discharge one of the pair of infra-red guided missiles. The work Saunders-Roe had already carried out led to an understanding that an improved derivative from SR.53 was possible, but chief designer Maurice Brennan also had doubts about the limited power available from the Viper engine, not to mention the lack of radar. He envisaged a much bigger aircraft, with a more rounded set of capabilities in the form of SR.177, and submitted his design to the Air Ministry. With interest from the RAF and Royal Navy, formal specifications were raised for the aircraft as

OR.337 RAF and NA.47 Navy respectively, with the RAF intending to operate the SR.177 in tandem with the English Electric Lightning.

For most of a sortie, power would have come from a reheated DH Gyron Junior DGJ.10-1 jet engine, then, should rapid climb be required, the latest version of the Spectre rocket motor fitted to SR.53 would have been engaged, giving climb figures in the region of 45,000ft/min and a top speed estimated at Mach 2.3. Equipped with an air intercept radar in the nose above a chin intake for the Gyron Junior and armed with a pair of *Firestreak* or *Red Top*, there was potential for a most impressive combat aircraft, offering rapid climb to height to counter any Soviet bomber threat.

At least nine foreign air forces had expressed an interest in the SR.177. Representatives from the West German government had visited Saunders-Roe, which offered the possibility of both the Bundesluftwaffe and Marineflieger operating the aircraft. Further afield, even the Japanese Air Force was said to be interested. With British and German interest, much was made about the potential for promoting the type to achieve commonality across NATO, a hot topic of the time, especially if NATO funding could be provided to support development costs. A potential order for 200 aircraft for West Germany had been discussed at ministerial level and enquiries had been made with Armstrong-Whitworth to facilitate licence production.

On 4 September 1956, the Ministry of Supply issued a formal contract to Saunders-Roe to supply 27 aircraft, with an initial order of nine development aircraft, split into three batches of three. This comprised a basic trio of general test aircraft, plus one batch each for the distinct versions envisaged for the RAF and RN, with the option to fit extended-range fuel tanks to the RAF aircraft dependent upon the sortie envisaged, while the RN aircraft would be fitted with arrester gear and high sink rate undercarriage for carrier landings.

The SR.177 order for the RAF was cancelled in the April 1957 Defence White Paper, despite five pre-production airframes already being in the later stages of construction. The Royal Navy requirement still existed, however, and work continued. Unfortunately, after political infighting between defence minister Duncan Sandys and the minister of supply. Aubrey Jones, who was willing to keep the naval version going, especially with proposed Luftwaffe/NATO funding, it all came to nothing, the contract for all UK aircraft being cancelled in August 1957. Meanwhile, the Germans looked on, bewildered that the UK could throw away a project at such an advanced stage of development. By December, they too had advised the UK government and Saunders-Roe that the SR.177 was no longer under consideration for purchase. On 24 December 1957, the Ministry of Supply officially informed the company that the aircraft was cancelled; happy Christmas. By the end of January 1958, all aircraft under construction were scrapped.

NATO funding never materialised and West Germany went on to buy Lockheed F-104 Starfighters from the USA, initially for the roles envisaged for the SR.177. A total of 916 was amassed by the Bundesluftwaffe, including the later F-104G multi-role versions, with the Japanese also buying 230 of Lockheed's aircraft. Bribery was later proved to have been involved in Lockheed selling so many F-104s to numerous NATO countries (Netherlands, Norway, Belgium, Spain, Greece, Turkey, Italy and Denmark all operated the F-104), along with other air arms around the world, which was all eventually revealed to the public in what became known as the 'Deal of the Century', leading to political fallout in a number of countries and the near collapse of Lockheed itself.

As for the British government's cancellation of an advanced project wanted by both UK services, with metal already cut, serial numbers allocated, and with the undoubted potential for export sales in the hundreds, that's another matter entirely.

Exhaust arrangements, with the Viper jet-pipe on top and the Spectre rocket exhaust beneath. The SR.177 reversed this arrangement, the reheated Gyron Junior needing a much larger jet pipe and intake than the Viper.

Not a missile, but a wire-reel flight test data recorder that could be carried by the SR.53. A far cry from the remote digital datalinks of today.

The de Havilland Spectre rocket engine fitted to the SR.53. This relatively compact unit developed a maximum of 8,000lb of thrust, though its appetite for fuel while doing so could consume the contents of the main fuel tanks in seven minutes.

XD145 with the Spectre rocket motor in full effect.

XD145 in flight; scaled up into the SR.177, it could have made a formidable warplane.

***Above and below*: Manufacturer's SR.177 models for both the Bundesluftwaffe and the Royal Navy.**

Chapter 8

Supermarine Type 510 – A Swift Saga

During the same period that Hawker was progressing from the Sea Hawk to the Hunter, Supermarine faced a similar challenge in designing a follow-up to its initial straight-wing jet model, but compared to its Hawker rival, it met with a very different level of success. The Supermarine Swift was ordered by a government eager to re-equip the RAF with aircraft able to outperform the MiG-15, and for a time, it appeared likely to enter service in advance of the Hunter. To a certain extent, however, the Swift was also ordered as an insurance policy should the Hunter not meet expectations.

Supermarine's first jet to see operational service was the straight-winged Attacker in 1950; it had first flown in 1946, but by December 1948, a more advanced version had already flown. This was the Type 510, in effect the marriage of a swept version of the laminar flow wing from the Spiteful, the final development from the Spitfire, with a fuselage built to accommodate a Rolls-Royce Nene jet engine. What it had also inherited from the Attacker and the Spiteful was a tail wheel, for which the Attacker had already been criticised, with any take-off from the many grass strips that were still in use resulting in deep grooves caused by the jet exhaust. However, the Type 510 at least attained the distinction of being the first British aircraft to have both swept wings and tail surfaces.

When evaluating the Type 510 at Boscombe Down, pilots were impressed by the roll rate and high speeds it was capable of; 670mph was recorded at one point, though they criticised the undercarriage arrangement again. It was suggested that switching to a tricycle format and improved elevator controls would result in a better fighter, with comparisons even being made against the F-86 Sabre, with which the Type 510 shared similar performance characteristics. Supermarine did address some of the above issues and also replaced the original needle nose of the 510 by fitting the more rounded nose from an Attacker, but it still retained the tail wheel. In this configuration, the Type 510 became the first swept-wing aircraft in the world to perform carrier operations, following a series of flights from HMS *Illustrious* in 1950, though Royal Navy interest in the design proved lukewarm. The aircraft was eventually assigned as a maintenance airframe and for a time, VV106 served as an instructional airframe at RAF Halton. VV106 is now held in storage at Cobham Hall, as part of the Fleet Air Arm Museum's reserve collection.

Progress towards what eventually emerged as the Swift was decidedly incremental, the next small evolutionary step being the Type 528, which flew in March 1950 and was built using the second Type 510 airframe (VV119). It incorporated a rare instance of reheated Nene engine. This aircraft returned to Supermarine after a very short period for yet more modifications and re-emerged as the Type 535, becoming airborne on 20 August of the same year. The sheer number of changes warranted a new type-number, but at least it finally gained a tricycle undercarriage as part of the process. Eventually, on 1 August 1951, the Rolls-Royce Avon-powered Type 541 arrived; WJ960 was the first true prototype Swift, followed by WJ965 as the second prototype in July the following year. This one was the first armed Swift, carrying a pair of 30mm Aden cannon.

The remaining Type 510 in Cobham Hall storage facility at the FAA Museum, RNAS Yeovilton.

The Swift became the last RAF aircraft to bear the Supermarine name, but unfortunately, after being pushed by the government to rush through the development, the Swift did not live up to its illustrious forebear, with only 197 produced of all marks, some of which never officially entered operational service. The Swift F.1 was plagued with poor flying qualities and engine problems with the early mark of Rolls-Royce Avon from day one. Climb to 40,000ft was deemed to be very good, but engine surging occurred at low air speed and high angles of attack, i.e. typical air-to-air combat conditions, which was less than ideal. The radius of turn was worse than the older Venom and the Sabre F.4, which by this time had been acquired for the RAF through NATO's Mutual Aid Defence Program (MDAP) as a stopgap pending arrival of the Hunter and the Swift. For a period, even the few Swifts that had reached the RAF were grounded, due to engine problems traced to a serious manufacturing fault with the turbine blades. Some of the more mundane-sounding things

The only surviving Supermarine Attacker F.1, WA473, on the mock carrier deck area at FAA Museum, RNAS Yeovilton.

that entry into service picked up undoubtedly point to an aircraft so rushed to get them into the RAF's hands, that they had simply not been tested or even considered, down to having a clear vision canopy with de-misting and a de-icing system that worked. The fitment of an extra pair of 30mm Aden cannon in addition to the original pair (a similar four-gun pack to that seen on the Hunter) resulted in the F.2, but this only made matters worse; centre-of-gravity changes caused by the weight and the adapting of the fuselage to accommodate the extra cannons plus their ammunition required another raft of modifications, not to mention further grounding due to a number of hydraulic problems.

Consequently, production numbers for early marks are very low, only 18 F.1s and 16 F.2s. The 25 F.3s built introduced a reheated Avon engine (that initially wouldn't work above 25,000ft!) but were never destined for RAF squadrons, merely declared as 'Instructional Aircraft'. The 62 F.4s built were the last purely fighter versions accepted by the RAF into service, retaining a now fully working reheated Avon and with a variable incidence tailplane fitted to finally overcome a long-standing pitching-up issue. But by this time,

Another view of the Type 510, essentially an Attacker fuselage with swept wings. Supermarine continued to use the tailwheel undercarriage despite previous criticism.

Type 510 VV106 in flight and with the original pointed nose.

Supermarine Type 535 VV119 under tow. This much-modified airframe began life as a Type 528 in 1950, but re-emerged as the Type 535 after modifications incorporating an afterburner, lengthened nose and tricycle undercarriage.

A flightline of Swift F.1s belonging to 56 (F) Squadron, the only unit to operate the type.

regardless of the best efforts of Supermarine and RAF engineers to fix faults and keep the aircraft flying, the Swift's days as a fighter in the RAF were over and based on the Chief of the Air Staff's recommendations, the Air Ministry grounded all Swifts on 15 March 1955.

That was not quite the end for the Swift in the RAF. The Gloster Meteors then in service and carrying out low-level reconnaissance tasks for RAF Germany were both obsolescent and increasingly reaching their fatigue limits, but with the urgent need to get the Hunter to the RAF as a fighter and ground-attack aircraft, production of these could not be diverted to fulfil another role. What was needed was an interim Meteor replacement and reasonably quickly.

Despite its limitations as a fighter and at higher altitudes, the robustly built Swift was duly modified for the photo-reconnaissance role and in FR.5 form went on to fill this gap before arrival of the FR.10 Hunter, going some way towards repairing a tarnished reputation. With the height restrictions imposed to achieve acceptable photographic images, the dense buffeting air demanded a strong airframe, and when used within a well-defined set of flight parameters, the FR.5 became a remarkably capable aircraft. It excelled in the low-level domain, winning the 1957 and 1959 NATO 'Royal Flush' reconnaissance competition against opposition from the likes of the RF-84F Thunderflash used by the USAF and a number of European air arms, and against RAF Canberras.

The Supermarine Swift is often disparaged when pitched next to the Hawker Hunter, but the comparison is slightly unfair. The Hunter also suffered from

Supermarine Swift FR.5 WK277 at Newark Air Museum. This aircraft is now at the South Wales Aviation Museum.

Swift F.4 prototype WK198, briefly a world record holder. The F.4 was the last fighter-only version accepted for RAF service. (Author's photograph courtesy Brooklands Museum)

Swept wings in the '50s. Swift FR.5 is on the left and Hunter F.1 on the right. Despite swapping the centrifugal-flow Nene found in the Type 510 and later gestations that led up to the Swift for the axial-flow Avon, the airframe seemed to retain the rotund midsection needed for the Nene, which was not satisfactorily designed out until the cancelled Type 545.

engine surge problems and hot gas ingestion when firing Aden cannons, with the engine issues only fully solved with a later mark of Avon fitted to the F.4. Much of the early pre-Swift development mentioned above was largely financed by Supermarine. Government funding arrived only once the Swift had been ordered, yet ordering production of 150 Swifts too early didn't actually help the aircraft, as it froze development to a great extent and as seen, well before a whole host of problems had been ironed out. The situation was compounded with only two official prototype Swifts, compared to ten similar Hunters built.

Beyond the FR.5, 14 special Swift F.7s were built with slightly extended wings, and a radar mounted in a longer nose, but these didn't enter operational service. Their role was to establish a dedicated guided-weapons development squadron, contributing a great deal of knowledge to the fledgling air-to-air missile programme when using the Fireflash missile.

A fully supersonic development from the Swift was in an advanced stage of construction as XA181 before the project was cancelled on 22 December 1955; this was the Type 545. It was powered by a reheated Avon engine and adopted a three-stage crescent wing shape derived from that of the short-lived HP.88

Swift F.7 XF114, the last remaining complete Swift F.7 under renovation. The 14 F.7s built were exclusively used for air-to-air guided weapons trials. (Author's photograph courtesy Solent Sky)

built for the Handley Page Victor development programme. The unfinished airframe survived for several years at the College of Aeronautics at Cranfield until it was eventually scrapped in the late 1960s.

Much like its Hunter contemporary, the Swift also held the world air speed record for a time, albeit briefly. WK198 was originally built as an F.1, but was later converted to an F.4, with a reheated Avon engine and variable incidence tailplane. Not content with setting a record London–Paris speed record of 669.15mph in May 1953 and despite suffering a forced landing at Supermarine's Chilbolton airfield in July following an engine seizure, WK198 was flown to Libya in September the same year, piloted by former Fleet Air Arm Lt Cdr Mike Lithgow[4] and on 25 September 1954 reached the then record speed of 737.7mph, eclipsing the three-week-old record of 727mph set by Neville Duke in the Mk.3 Hunter. The record wasn't held very long, however, as an American Douglas Skyray pushed the record to 753mph only a matter of days later.

As XF114 would have looked when part of the Guided Weapon Development squadron, complete with a pair of underslung Fireflash beam-riding missiles.

4. Mike Lithgow had been Supermarine's chief test pilot since 1948. Sadly he perished with six other BAC flight crew, when the BAC One-Eleven prototype G-ASHG crashed on 22 October 1963.

Chapter 9

Lightning Legacy

If you aim to design an aircraft advanced enough to remain in service for more than 35 years and still be the fastest all-British production aircraft, then you must ensure that this design behaves across the whole of the flight envelope before production commences. The aircraft in question is, of course, the English Electric Lightning, designed by 'Teddy' Petter and Freddie Page, and conceived by Petter as a development from a proposal for a single-engine research aircraft as far back as 1947. The research aircraft eventually became Project 1 or P.1, a radically different aircraft compared to any of its contemporaries. It had highly swept-back wings, essentially a delta with notches cut out of the trailing edges, plus matching tailplanes and two powerful engines stacked on top of each other with the smallest practical airframe wrapped round them. Fine in theory, but the airframe size, or rather the lack of space within it, to store fuel or provide space for other

WG760, the first P.1A prototype, at the RAF Museum Midlands.

equipment, unwittingly restricted the Lightning's development potential further down the line.

Short SB.5

To achieve the Mach 1.5 originally desired, it was already understood that highly swept-back wings would be needed, but little was known about low-speed flight in such a configuration. Therefore, a thoroughly original development aircraft was built to address this lack of knowledge; this was Short SB.5 WG768.

Capable of having the wings set at three different sweep angles (50, 60 and 69 degrees) as well as a changeable tail section allowing either a T-tail or low-mounted tailplane configuration, this design represented value for money. The tailplane position issue had resulted from a difference of opinion between English Electric (EE) and the Royal Aircraft Establishment (RAE) over the proposed wing planform, not to mention the tail position. The RAE suggested that the T-tail would be the optimum fitment, whereas EE maintained that the low mounting would be better. The RAE was ultimately proved wrong on both counts, as the P.1 and the Lightning kept English Electric's original low-mounted tail surfaces.

The low-speed environment allowed the SB.5's wing structure to be built largely of plywood, with alloy leading-edge panels, while the rest of the airframe was conventional stressed-skin alloy. The wooden wings in turn limited the top speed to 350kts and removed any need for a retractable undercarriage, reducing weight and complexity. The SB.5 was first flown on 2 December 1952, with a 50-degree wing sweep and the T-tail unit in place. By July 1953, the 60-degree sweep wing had been adopted, and the T-tail gave way to the low-set pattern on 10 January 1954. The development programme continued for four years until 1958, when the aircraft underwent an engine change to rectify a major issue from day one; it was drastically underpowered. The 3,150lb-thrust Rolls-Royce Derwent thus gave way to a 4,850lb-thrust Bristol Siddeley Orpheus. The wing setting was changed to 69 degrees at this time, and the final set of flight trials began on 18 October 1960 from RAE Bedford. All the test data gathered during the various stages of the SB.5 trials was fed into the P.1 project and the Lightning development programme.

The SB.5 proved to be such a versatile machine that the story didn't end there. It carried on providing valuable input into other projects, including Concorde from 1964, with the port wing at one time fitted with a series of 'tufts' to demonstrate leading-edge airflow patterns. Finally put into semi-retirement at the Empire Test Pilots School, WG768 eventually ceased flying in 1967 and

Above: The only Short SB.5 built, WG768. This is now on display at the Ulster Aviation Collection.

Left: The SB.5's tail section in the 'normal' configuration with the tailplanes set low. Note the hatch for the drogue chute above the jet pipe.

Opposite above: Spare Tail: The T-Tail section built for the SB.5 at the insistence of the Royal Aircraft Establishment.

Opposite below: SB.5 in flight, with the T-Tail fitted.

WG768
HIGH TAILPLANE FUSELAGE

WG768

English Electric P.1 WG763, the second P.1A prototype built, is seen here at Manchester Museum of Science and Technology. It is now at the Boscombe Down Aviation collection, Old Sarum, Wiltshire. A third P.1A airframe was built, but only as a static test aircraft.

until December 2025 stood beside the P.1 at RAF Museum Midland, compete with the optional T-tail rear fuselage section beside it for comparison (all the photographs were taken while it was still at Cosford). WG768 is now on display at the Ulster Aviation Society's museum at Long Kesh in Northern Ireland.

English Electric P.1

English Electric received the contract to build the P.1 on 1 April 1950, with three airframes ordered, two flying and one for stress testing. A further three prototypes were later ordered, and the original flying pair became P.1A models, the second batch as P.1B. Delays to the Rolls-Royce Avon engines for the production machines resulted in the P.1A being powered by Bristol Siddeley Sapphire engines.

The first P.1A, WG760, flew on 4 August 1954 and after only three flights became the first British aircraft to break the sound barrier in level flight (and this without a reheat unit, yet to be fitted). Once reheat became available, the rate of climb increased to near 11,500ft per minute. Due to a directional stability issue found to be caused by the fin being too small, the P.1A was later restricted to Mach 1.5. The P.1B rectified the fin problem and finally incorporated a pair of Avon R24Rs, while English Electric had designed a new shock cone to be mounted in the centre of the air intake, allowing fitment of AI (Air Intercept) Mk.23 radar, which in the process gave the P.1B a look considerably more akin the final production Lightning. The changed air intake and increased fin area also allowed English Electric to inform the Ministry of Supply that it expected that the P.1B would be able to exceed Mach 2. Regardless of doubts at the ministry and the customary 'no funds for this' noises, specification F.23/49 was amended to allow the building of XA847. This first flew on 4 April 1957 and eventually, after 258 flights, with each sortie extending and checking the aircraft's flight envelope, exceeded Mach 2.0 in level flight on 25 November 1958. XA847 still survives and is now at RAF Manston History Museum.

Despite being faster than the Fairey Delta 2, which had attained the world air speed record of 1,132mph on 10 March 1956, the P.1B lacked sufficient fuel capacity to complete the two runs demanded by international rules.

With 337 Lightnings built of all marks, including prototypes, the only RAF aircraft to emerge unscathed from the 1957 Defence White Paper debacle went on to serve until 1988. If not for the fallout following the White Paper, the fruits of further development of the Lightning airframe may have been seen; even a proposed naval version using a variable-geometry wing was suggested, but none progressed off the drawing board.

P.1B XA847, the first of three built and the only survivor. This aircraft was the first British aircraft to achieve Mach 2.0 in level flight in the hands of Roland Beamont. Held in private storage for many years, it can now be seen at RAF Manston History Museum in Kent.

XG337, the last pre-production Lightning F.1 aircraft, at the RAF Museum Midlands. Hung in what was to become a characteristic Lightning 'pose', it illustrates the unique wing and tail planform.

Norfolk and Suffolk Aviation Museum's Lightning F.1, XG329. Built as part of the F.1 development batch in 1959, this initially went on to conduct supersonic cannon firing trials and *Red Top* missile development. It was later used as a test aircraft for the Lightning F.3, hence that version's larger flat-top tail, needed for fitment of *Red Top*. XG329 also acted as a chase aircraft during TSR-2 development flights.

Generation gap: English Electric P.1A WG763 sits next to a Griffon-engined Spitfire at the English Electric site at Warton.

Chapter 10

TSR.2

'All modern aircraft have four dimensions: span, length, height and politics. TSR.2 simply got the first three right.'

Sir Sydney Camm

Of all the aircraft that made it off the drawing board and into the air but not into operational service during the 'Golden Age' of the 1950s and early '60s, none evoke as much passion, comment and opinion as TSR.2. Despite more than half a century having passed since its cancellation in April 1965, to many there remains an element of conspiracy theory surrounding its demise; questions unanswered by succeeding governments, or stories not hinted at in any minister's or senior RAF officer's memoirs. If you gather any number of British aviation enthusiasts together and mention TSR.2, you could be there for days!

When originally conceived, the project was purely as a supersonic replacement for the Canberra in the light strike and reconnaissance roles, set out in GOR.339 (General Operating Requirement) in 1956, initially with a brief that it should operate at very low level, attain Mach 1.5 at altitude and possess a 2,000-mile ferry range. While still a paper project, GOR.339 managed to sneak past Duncan Sandys' 1957 Defence White Paper, despite the 'manned aircraft are obsolete' rhetoric, while others in far more advanced stages of development, or even pre-production, perished. Once saner heads prevailed regarding missiles over aircraft, and aircraft companies in the UK attempted to make up for several wasted and unproductive years, further discussions within the government led GOR.339 finally being opened to tender across the aircraft industry. In the process, it was used as something akin to a cudgel by the government, bludgeoning the industry into reorganising itself, purely in order to secure work on the contract. This eventually saw the joint offering by Vickers and English Electric being selected and resulted in the pair becoming the first elements of BAC, the British Aircraft Corporation.

The Australian Air Force was initially interested in TSR.2 but opted for the F-111, yet it had to wait a further ten years before finally taking delivery of the first one into service, and by then was faced with a contract which had trebled in cost over that decade. Ultimately, it is possible that both the

XR222, what would have been the fourth prototype, at IWM Duxford's Airspace Hangar.

RAAF and the RAF could have had TSR.2 sooner and potentially cheaper, but the Australians had also wanted to move closer to the US in terms of strategic defence, and without a concrete RAF order for TSR.2, it was always a non-starter. Comments attributed to the then head of the Royal Navy, Lord Louis Mountbatten, to senior Australian service personnel about the RAAF being able to buy five Buccaneers for the cost of a single TSR.2 and that it would never be built anyway, didn't help either. Meanwhile the government's chief scientific adviser, Sir Solly Zuckerman, decried the aircraft and the UK aircraft industry in general at every opportunity; though it is debatable whether a zoologist such as him truly understood either subject.

In comparison to some of the other cancelled projects, TSR.2 was the closest to making the final hurdle from prototype to production and then squadron service. So certain was the RAF that TSR.2 would arrive, that engineering personnel were undergoing training courses to support the new aircraft (a situation later repeated during the failed Nimrod AEW project in the 1980s). TSR.2 had money lavished on it, around £160m at the time of its demise (£4.02bn in 2026 figures) and it was James Callaghan, then chancellor, not Denis Healy, the Labour defence minister, who finally swung the axe, estimating that he would save the taxpayer £700m (£7.9bn in 2026 figures) over a 15-year period by cancelling it. Despite the cost of upgrades that brought into service the Hawker Siddeley Buccaneer, which eventually filled the strike mission intended for TSR.2, plus some part of the cost paid for UK-specification Phantoms that took on the reconnaissance role, it is hard to see where this figure came from, as the combined costs for these developments would have left a huge amount remaining from that total.

TSR.2 Timeline

Year	1958	1959	1960	1961	1962	1963	1964	1965
Estimated Total Project Cost.	M£137	M£211	M£285	M£359	M£433	M£587	M£741	M£750
Events	R&D Costs expected to be M£35. M£80-90 including 9 development aircraft.	First suggestion to cancel by Treasury due to rising R&D costs, estimated to have doubled to M£70	Oct - Initial development batch ordered. R&D estimate - M£90	Nov - All electronics already significantly lagging airframe development	Mid-year R&D costs expected to be M£137. Rose to M£175 by year end.	R&D Costs including engines now expected to be M£175-200	Mid-year R&D costs estimated M£240. M£272 by October.	Last test flight - 31st March.
			Strategic bombing requirement added to spec in 4th Draft		Blue Water Missile cancelled - Funding split between TSR.2 and Polaris	Oct 29th - first public press release.	First Flight - 27th September	Cancelled - 6th April
					Vulcan B1 XA894 Olympus 22R engine test bed destroyed.	June 12th - 11 pre-production a/c ordered to secure long lead time materials.	October 14th - UK General Election	Actual project costs to date - M£160
		First flight Expected - 1963	NA.39 development and F-111 ruled out as an option for GOR.339			Original development batch 2 months behind schedule.	Ordering 158 F-111 estimated to save M£250.	Options placed on F-111A from USA. Cancelled Jan 1968.
Estimated In Service date	*Late 1963*	*1965*	*1966*	*Late 1966*	*Late 1967*	*End 1968*	*1969*	*1970*
Cost Per Aircraft - 150 Ordered		M£1.7			M£2	M£2	M£3	M£5
100 Ordered								M£6
Notes	NA.39 - Blackburn Buccaneer S.1. By the time of cancellation, TSR.2 was expected to cost more than new CVA-01 aircraft carrier project for the RN (also later cancelled) and approaching half the desired M£2000 defence budget. The chart behind is an indication of how costs escalated out of control, with overall project figures (top line) racing ahead of R&D estimates (lower), while the in service date slipped year-on-year.							

Much of the anger directed at the Labour government of the day following the cancellation was not so much in finally terminating the project in April 1965 (rumours having circulated in the press for months, even before the 1964 elections ousted the previous Conservative government) but in the manner which it was carried out. Only one aircraft flew, XR219, but this, along with XR221 and XR223 (only part-completed) was taken to Shoeburyness range and used for target practice to 'determine the effect of gunfire on modern aircraft structures'. (Questions regarding the suitability and durability of the materials choices for TSR.2 circled for a long while.) Yet the second prototype XR221 was complete and scheduled for a first flight on the day of cancellation, but orders from the government forbade this from happening, though the flight crew threatened to take it up anyway. Three further fully completed airframes, XR224, XR225, XR226, along with another part-completed example, XR227, were simply scrapped. Only XR222, which was virtually complete, along with XR220, which was in a far less advanced state, survived the dismantler's blowtorch or gunfire, but only just. Worse still, in an act of apparent government-sponsored vandalism that seems to raise the most ire, the manufacturing jigs were also scrapped, thus stopping BAC from even thinking about a private venture (as turns out, this was the normal practice, largely because there is insufficient space to store them). This also prevented any chance of attempting to resurrect the project three years later, when the orders for the 'cheaper' replacement, the General Dynamics F-111K, were themselves cancelled, due to delays in that aircraft's development and rapidly spiralling costs – problems that sound rather familiar.

Where did it all go wrong? Sir Sydney Camm summed the biggest issue up perfectly in his comment that heads this chapter; TSR.2 could well have been a perfectly good aircraft, if not an outstanding one, had politics not killed it. The trouble was that it wasn't only the politicians playing politics. The more you read into the sorry tale of inter-service rivalry and corporate half-truths, the more you realise that the project was ultimately doomed from the start. The great pity is that all this took place in the public eye and cost millions from annual defence budgets that could have been spent on other much-needed projects.

There is no escaping the fact that delays in the development programme and inexorable rising costs gave ammunition to those wishing TSR.2's demise. Hindering progress were problems with the complex undercarriage mechanism, which displayed a stubborn unwillingness to retract, while vibrations traced to the fuel pumps were at such a frequency that they affected the pilot's vision. Delays in engine development that dogged the UK aircraft industry post-war also impacted TSR.2. Work on the Olympus 22R engines were not helped by a

Vulcan B1 XA894, with an Olympus 22R testbed on the centreline. Total loss of this aircraft in a ground fire, when the 22R on test disintegrated, greatly set back engine development.

ground fire destroying the Avro Vulcan B1 used as the testbed aircraft, where by a development engine was mounted in a nacelle under the fuselage; the cause was later attributed to the disintegration of turbine blades caused by harmonic resonance. Unbeknown to XR219's flight crew, should maximum power on one engine have been needed on the first flight due to an emergency, there was every chance the same catastrophic engine failure could have been repeated. This turbine fault also led to a three-month hiatus between the first and second test flights, potentially the most fatal delay of the whole programme, but maybe only to the extent that it would have been Conservative defence minister Peter Thornycroft who cancelled the project prior to the 1964 election, rather than the Labour government that followed.

Despite all these tribulations, once the Olympus was working well enough for full flying trials, XR219 on its 14th flight went supersonic over the Irish Sea on 22 February 1965. Test pilot Roland Beamont's post-flight quote showed his confidence in the aircraft: 'It handles like a fighter at supersonic speeds and in turbulence is as steady as a rock. It is a fascinating aircraft to fly and, in some ways, the finest I have ever flown.'[5] Little did he know, the project had less than two months before cancellation.

Most debate about TSR.2s concerns the great what-if's that the test programme never got the chance to address. For example, XR219 never flew

5. *Air Pictorial*, April 1965.

with a full set of avionics, where features demanded by the final iteration of OR.339 really needed technological advances in electronics, which wouldn't arrive until several years later, certainly well after the point when TSR.2 would have been in service. Even adding more mission profiles to the specification late in the day had an impact, especially on the limited computing power available at the time, while the viability of reliably uploading these profiles to the aircraft pre-flight, via paper tape (say on some wind- and rain-swept airfield, in the middle of the night and in a hurry) is somewhat questionable. The avionics processing system itself was provided by a pair of Verdan flight computers derived from the single unit fitted to the North America A-5 Vigilante used by the US Navy and adapted for TSR.2. Future development costs here could certainly be where a good deal of the £700m savings might eventually have come from, but it was doubtful that this could even have been guessed at, let alone calculated. The Decca terrain-following radar (TFR) feeding the inertial navigation system (INS) that was being built by Ferranti was another potential money pit, and one that could have partly been avoided. An existing INS system produced by Litton Industries and working well in more than 1,000 F-104 Starfighters was already demonstrating better accuracy than the one being built by Ferranti for TSR.2. This product had been suggested during the early stages of the project, but the Ministry of Aircraft would have none of it. Despite TFR not being fitted, XR219 attained high subsonic speeds at 200ft. Had TFR been installed, who knows how fast it could have flown at the truly low levels for which the Buccaneer became renowned.

TSR.2 started to gain a nuclear-armed strategic bomber element to its roles well before it even flew, in signs that senior RAF staff were trying to maintain some sort of hold on Britain's strategic deterrent before the Navy gained Polaris. A scheme to mount one or two modified versions of the 'Blue Water'[6] theatre tactical nuclear missiles on TSR.2 had been proposed by BAC in 1962. This followed the demise of the American Skybolt, which was intended to equip the Vulcan. Air-launched Blue Water would have provided around a 90-mile stand-off range, with 1,500-mile radius of action for TSR.2 when so fitted, including a radar-evading low-level attack profile in the later stages of a mission. This and other suggestions did little to speed progress of what started out simply as a supersonic Canberra replacement, while the requirements for the low-level attack profile added millions to the

6. Ironically, Blue Water for the Army had been cancelled in August 1962, after £20m had been spent on it, with the £50m budget previously allocated to continue development, used for TSR.2 and Polaris. This decision also cost 2,000 jobs after the closure of English Electric's guided weapons department.

XR219 awaits another test flight. The TSR.2's size is deceptive; certainly larger than you would expect at nearly 6m longer than an Avro Lancaster, just over 7m longer than a Canberra and more than five tons heavier than both.

development costs. Likewise, demands for exceptional short field take-off performance, originally including a proposal for a custom lift jet-powered launch platform built by Short Brothers, again added to costs and overall build complexity.

Following the cancellation of P.1154 and HS.681 in February 1965, TSR.2 only survived a few weeks longer, largely because of the media attention it had already attracted. The new government also considered that announcing all three cancellations at the same time would be too much bad press all at once. After all, in the lead-up to to being elected, future prime minister Harold Wilson trumpeted that if Britain was to prosper, it would need to do so 'in the white heat of technology'.

TSR.2's demise has often been accompanied by intimations that financial pressure was applied on the UK government by the US to curtail the project. This was ostensibly to ensure that the F-111 programme continued without comparison to TSR.2, even to the point of the USAF's cancelling F-111 in favour of TSR.2. It is plausible, but what wasn't made public until some years later was that in the early 1960s, the US had offered the UK government the F-111, F-4 Phantom and C-130 Hercules, plus industrial offsets over a 14-year term at just over 5 per cent interest. This would have avoided a potentially huge projected peak in the defence budget, to a point where if development continued and costs kept rising, work on TSR.2 amounted to virtually the entire RAF annual budget. Orders for those three US aircraft did eventually replace those which could have been for TSR.2, P.1154 and HS.681. Only the

F-111's own costly and protracted development, alongside Britain's worsening financial situation, failed to see it fly with RAF roundels.

Just 24 test flights were carried out using XR219, but feedback from the flying programme indicated that the aircraft possessed wonderful flying characteristics. Chief test pilot Roland Beamont said at the time that after the initial bugs had been ironed out of the undercarriage and engines, 'he was looking forward to two years of development flying in a highly capable aircraft'. The airframe itself might have been superb, but the stringent requirements of OR.339 probably could not be met in full, and it's here where the true crux of TSR.2's ultimate demise arose; the original requirement itself. If the lily had been a little less gilded, and it continued to be even late in the day, the resulting aircraft may well have achieved service and could well have earned the country valuable overseas sales. The final requirement specified Mach 2+, but if the Mach 1.5 or 1.7 of the first draft had been retained, it would still have been twice as fast as the Canberra or any V-Bomber and significantly, there would have been no need for the troublesome Olympus 22R engine. The extra thrust required from these and the additional airframe heating they created pushed the aircraft beyond Mach 1.7, just like the Bristol Type 188, meaning more expense, special alloys, more fuel, more cooling, more testing and crucially, more delays.

Non-existent overall project management essentially crucified the TSR.2 programme, and this is generally accepted as another root cause for the cancellation. A poor government decision at the outset that gave the project lead to Vickers, when the actual aircraft was based on EE's P.17A designs, and to which the final aircraft bears a distinct resemblance, simply caused unnecessary friction between the design teams. There were further arguments over who built which sections and where final construction would be. Then there were the government department-mandated decisions as to which systems to fit and which were to be built by whom, despite the design team stating their preferred suppliers, even down to Bristol Siddeley rather than Rolls-Royce to supply the engines. Crucially, there was no single leadership team or design authority holding all the information, from all the companies involved. It was often the case that some of the firms with systems destined for TSR.2 were at times 'economical with the truth', praying that another party would have to officially announce an issue and cause a delay, so giving them a breathing space to cure their own problems on the quiet. On one occasion, when an all-parties meeting was eventually convened, 60 people turned up; the chairman dismissed them and asked them to reconvene on a later date, once it had been established who honestly needed to attend. However, 65 people arrived for the next meeting! The list of what went wrong and why simply goes on and on.

To the aircraft's eternal credit, the failure of the TSR.2 project and the reasons that contributed to its failure actually benefited every other military aircraft programme that followed it, though contrary evidence would suggest that is debatable at times.

XR220 and XR222 now sit in their respective museums, beautiful white tributes to what might have been.

A Mirage IV for the RAF? With the demise of TSR-2, BAC went as far as suggesting a joint project with Dassault to produce a Rolls-Royce Spey-engined version of the French delta-winged bomber, whose size and weight were comparable to TSR-2, but it lacked the range and low level abilities. The only Mirage IV on display outside France stands in the autumnal damp. (Author's photograph, courtesy Yorkshire Air Museum)

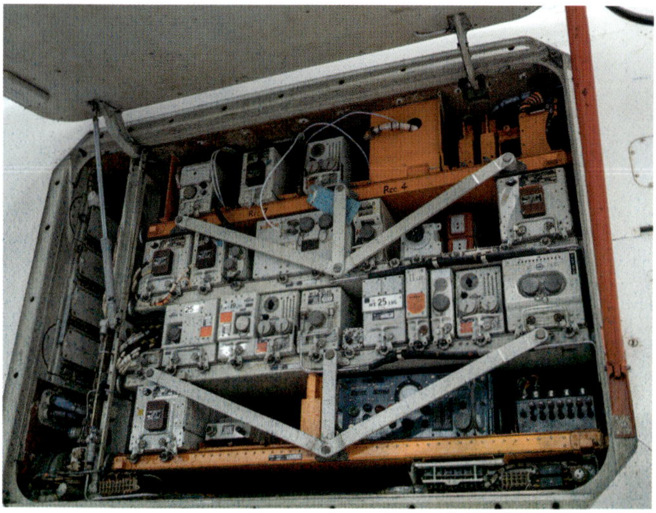

The TSR-2 equipment bay, where some thought appears to have gone into making it accessible for servicing. The large access panel provides overhead protection during bad weather and a set of covers could be attached as side screens.

The problematic undercarriage. It took ten test flights before the undercarriage could be retracted. The thin hydraulic strut behind the main leg here finally solved the serious vibration issues.

RAF Museum Midlands TSR-2 XR220 is impressive regardless of its age. It has a 20ft-long internal weapons bay designed to accommodate up to 6,000lb of ordnance, originally including a single *Red Beard*; or later could have taken a pair of WE.177s, tactical nuclear weapons or a dedicated reconnaissance pod, plus wing hardpoints for a further 4,000lb of external stores, fuel tanks and so on.

How does one attach the fin and tailplanes on an aircraft that is expected to be buffeted in the dense air of low altitudes? This is the mounting frame from an undetermined TSR-2 airframe located at the Norfolk and Suffolk Aviation Museum.

Not a cockpit from one of the airframes broken up, but a specially built unit for thermal testing. (Author's photograph, courtesy Brooklands Museum)

***Above and below*:** **RAF Museum Midlands' cutaway Bristol Siddeley Olympus 22R is big enough on its own but, when attached to the lengthy jet pipe and reheat unit, it seems enormous. This also illustrates just how far the main engines themselves were buried inside the fuselage.**

Above: We are exceptionally lucky to have two examples of TSR-2 still surviving, as the more the matter is researched, it becomes very apparent that the government of the day would quite happily have quietly buried the remains of the whole project. XR222 is at Duxford before it was placed within the then newly refurbished Hangar 1.

Left: XR220 stands proud in the centre of what was the Test Flight Hall at RAF Museum Midlands. So many 'what ifs'.

Chapter 11

Fairey Delta 2

Fairey's building of delta-winged research aircraft began with its submission for a ramp-launched interceptor to meet Air Ministry Specification E.10/47, and in the process this became the first British aircraft with a delta wing. After evaluating numerous designs, the Ministry of Supply finally decided to build one to investigate delta-wing flight characteristics and controls. Based around a 3,600lb-thrust Rolls-Royce Derwent engine, the resulting small and rather rotund airframe utilised a 19ft 6½in delta wing and became the Fairey Delta, making a first flight at Boscombe Down on 12 March 1951. It was found to have some serious stability problems and despite adopting a small delta T-tail to counter a pitching issue as it gathered speed, it was still rated as dangerous to fly. Government funding ceased, as did its interest in ramp-launched fighters, but Fairey continued to use the aircraft for experimental purposes until a landing accident in 1956 rendered it beyond repair, after which it finished its days as a gunnery target at Shoeburyness and was later scrapped.

Fairey's original delta-winged research aircraft, the FD.1, which was built at its Heaton Chapel factory near Stockport. The rather portly fuselage around a Rolls Royce Derwent engine giving a less elegant look compared to the FD.2 which followed.

Fairey Delta 2 WG777 looking resplendent in gloss blue finish at RAF Museum Midlands.

As it became increasingly obvious that more data was needed on aircraft control at transonic speeds, possibly with an inkling towards requirements shaping up for aircraft to follow on from the Hunter and Javelin, a new specification, ER.103, emerged. Fairey decided to build a machine vastly different in stature and performance to its predecessor, though its naming as Delta 2 was less imaginative and this quickly became shortened to FD.2. With the Fairey Gannet anti-submarine warfare (ASW) aircraft classed as a 'super priority' programme at that time, work on the FD.2 was held up, with construction not starting until the summer of 1952. First flown on 6 October 1954 by Fairey chief test pilot Peter Twiss, the aircraft went supersonic without the need for reheat on 28 October 1955 and Fairey soon realised that there was considerable potential in the design. The FD.2 became a highly successful research aircraft and went on to set a new world air speed record of 1,132mph on 10 March 1956.

Powered by a single Rolls-Royce Avon 14R engine with 9,500lb thrust normally and 14,500lb with reheat engaged, the dart-shaped aircraft, with the then unique 'droop snoot', which provided improved pilot vision on landing and taxiing, was superb from the outset. The high speeds achieved were way beyond those anticipated by the Ministry of Supply; even Rolls-Royce doubted the viability of its engine beyond Mach 1.5 and had the temerity to criticise the air inlet shape before being made to eat its words.

During the test-flying programme, the story became interesting for other reasons. Following the world record run and a new set of government rules that severely limited supersonic flying, certainly over land in the UK, the flying schedule moved to Cazaux in south-west France, at a facility owned by Dassault, with whom Fairey had a good relationship. Over 40 flights were carried out from Cazaux, during which it was entirely probable that Dassault observers learned a good deal about delta-winged aircraft. Not long after the FD.2 departed, Dassault brought to completion a design off its drawing boards that bore a strong resemblance to the FD.2, but originating before anything learned from the FD.2's time at Cazaux could seriously influence the project. Over 1,400 aircraft later, the Mirage III and its variants served with 14 different air arms across the world and acquired an exceptional combat record in the hands of the Israeli Air Force.

Fairey itself had plans for a two-seat fighter, benefiting greatly from information garnered during FD.2 flying and had submitted the design to contend for RAF specification OR 339/F.155T to build a missile-armed interceptor. Based on a near double-sized FD.2 planform, it retained the droop snoot but swapped the single Avon engine for the considerably more powerful de Havilland Gyron or Rolls-Royce RB.122; it was to be armed with a pair of *Red Dean* or *Red Hebe* missiles. Unfortunately, the Fairey ER 103C (or 'Fairey Large') was consigned to the scrapheap by the 1957 Defence White Paper, along with virtually every other aircraft on company drawing boards.

However, Fairey also had a more modest upscaling of the FD.2 on its boards; naturally christened 'Fairey Small', it could conceivably have been developed into a capable combat machine. With the FD.2 amounting to a flying prototype already built, Fairey estimated that it would only take 18 months from an order being placed to first flight. This would have given the RAF a Mirage-class aircraft by 1960, not to mention potential export orders for Fairey. Comparing the FD.2 to the Mirage, it must be remembered that the original Mirage prototype was more akin to the slim FD.2 than the eventual bulkier-looking production aircraft, which had grown slightly to allow the radar and weapons mounts missing from the original prototype.

The original FD.2 aircraft would never have made a combat aircraft and was never designed to be one. The considerably larger ER.103C was a stage too far, reliant as it was on an engine that never flew unless it was in a housing belonging to another aircraft (the Short Sperrin), while the missile that was its intended armament had even more development problems, though alternatives would soon be available. Even so, adopting a more modest approach and building 'Fairey Small' would have made far greater sense, as the potential undoubtedly existed within the basic FD.2 shape for a successful production aircraft. But without a government contract and hence substantial funding, Fairey sadly failed to pursue this; Dassault readily admitted that it could never understand why.

If you superimpose the FD.2 planform over that of the Mirage, it bears an uncanny similarity to Dassault's finest, especially the Mirage prototype, which was slimmer than the eventual operational aircraft.

Mirage III. A British-designed and built competitor to Dassault's highly successful fighter, derived from the FD.2, was eminently feasible. (Author's photograph, courtesy Yorkshire Air Museum)

The Fairey ER.103C model, complete with either *Red Dean* or *Red Hebe* on its wingtips, was a considerable scaling up of the original FD.2 design.

WG774 in flight prior to conversion into the BAC 221 and with the nose attitude set for landing.

Chapter 12

Testing for Concorde

Even before the British and French governments agreed on a joint programme to develop the supersonic passenger aircraft that became Concorde, several design projects and investigations had already been carried out on both sides of the channel.

Research at the Royal Aircraft Establishment, primarily led by Johanna Weber and Dietrich Küchemann, had concluded that a highly swept delta wing that ran for nearly the whole length of the aircraft's fuselage could meet all the requirements for both high-speed and low-speed flight. At low speeds, as much lift as possible is needed just to get airborne, but once speed increases and especially at supersonic, changes in airflow and compressibility require a slender, short-span wing. The 'Ogee' wing, as it was named, was found to generate an airflow vortex at low speed that provided the required lift, while its short span also provided the optimum planform once into the supersonic domain. What was needed now was an aircraft to test how well this could be controlled at both low speeds and high speeds, and these requirements were furnished by the HP 115 and BAC 221.

HP.115

The original plan to explore the low-speed flight regime was to build a glider to be towed aloft by a Canberra. Construction was to be the responsibility of UK glider manufacturer Slingsby Sailplane, but the economics and practicalities of this method were soon dismissed as inefficient, and development was stopped. Instead, a powered version was proposed and tenders let out; Handley Page was eventually awarded the contract and the slightly awkward looking HP.115, XP841, was the result.

With a Bristol Siddeley Viper engine providing 1,900lb of thrust mounted at the base of the tail, the centre section of the fuselage looked rather thin, leaving the forward fuselage appearing slightly bulbous as a result; this was purely to accommodate the pilot and his seat. The wing blended in to virtually the full length of the fuselage, starting as the cockpit finished. The HP115 did not, however, have a true Ogee wing, but a relatively simple, long slender delta, which was all that was required for slow-speed testing.

HP.115 head on, showing the very short span wing.

By the time XP841 first flew on 17 August 1961, the British government had already put pen to paper with its French counterparts, so all the test flying carried out by the HP.115 fed directly into the Concorde programme, though HP.115 flights ultimately ended in 1974, well after Concorde itself had flown.

BAC 221

The testing of designs for Concorde in the supersonic domain led Fairey to suggest that it could modify the FD.2 airframe to provide an aircraft closely resembling that proposed for Concorde for the high-speed regime. However, with Fairey now part of Westland, it was decided to hand the work to Bristol and the second FD.2 built, WG774, was flown to this company's Filton factory in September 1960.

Following the merry-go-round of enforced company mergers, the aircraft eventually appeared as the BAC 221 on 7 July 1961. Among its features was a lengthened fuselage, while the former wing root intakes had been replaced by rounded inlets under new Concorde-like wings. There was also a scaled version of the tail fin, as well as a lengthened nose landing gear to simulate that expected to be fitted to Concorde, so as to study the expected take-off and landing characteristics.

After two years of development flights, WG774 was fitted with a more powerful Avon RA28 engine, while further modifications were made to the air inlets. These and other changes restricted the top speed of the 221 when compared to the original FD.2 design (Mach 1.6 as opposed to Mach 2.1), but one benefit was increased fuel capacity, which allowed longer test flights.

Dwarfed by the trailing edge of Concorde prototype 002, XP841 resides in a corner of the RNAS Museum at Yeovilton. The domed pod near the top of the tail is a camera housing used to film smoke flow over the wings.

XP841 in flight in 1962.

In total, the BAC 221 provided ten years of test flights, all providing valuable data input into the Concorde programme, before its retirement in 1973. Rightfully preserved after providing such a valuable service, WG774 sits alongside Concorde 002, the first British prototype, at the Fleet Air Arm Museum, Yeovilton.

Smoke On! An unusual view of XP841.

BAC 221 WG774, on an early test flight. One set of front undercarriage doors appears to be missing, while the forward pair are open.

Left: Close-up of the air inlets on the BAC 221 and the long forward fillet of the wing fuselage join that characterised the wing of Concorde.

Below: The BAC 221 at Yeovilton, surrounded by a Concorde 002, Hawker P.1127, BAe Sea Harrier and a Westland Wyvern.

Above and below: **Concorde G-BBDG 'Delta Golf', showing the blending of the wing and fuselage, as well as the long forward wing section of the Ogee shape. (Author's photographs, courtesy Brooklands Museum)**

In formation, BAC 221 WG774 and FD.2 WG777 form up for a unique shot that highlights the changes to the wings, especially between the original FD.2 and the BAC 221.

Chapter 13

Red Dean

Referenced on several occasions alongside the aircraft intended to carry it, examples of the Vickers *Red Dean* missile are on display in several museums. Seeing them up close underlines the problems that aircraft designers had in trying to facilitate their carriage, while also illustrating the limitations of the technology available at the time. Designing an aircraft to fly at Mach 2, with a pilot aboard, sufficient fuel to give it useful range and a radar bolted in the front to find its intended targets is one thing, and it would be expected that the companies involved had the experience needed to achieve this. Designing a guided weapon to be launched by the same aircraft, to hit another aircraft potentially at a similar speed, at an adequate range, while carrying a warhead big enough to destroy the target or at least cause sufficient damage to let gravity do the rest, when still using valve-era technology, was another matter entirely. It was also a subject where at the time, concrete expertise was decidedly lacking.

Starting in 1947, the Ministry of Supply issued a series of operational requirements for air-to-air guided missiles (AAMs) with OR.1056 for Red Hawk, a missile that would later be called an all-aspect guided weapon, but by June 1951, it was realised that this engagement envelope was beyond the technical capabilities of the time and the requirement was dropped. However, 1951 also saw the issue of a less demanding requirement, OR.1117. This was allocated a Rainbow Code of *Blue Jay*, which would eventually become the infra-red homing *Firestreak*, often seen carried by Lightning, Javelin, and Sea Vixen. OR.1105 was issued the same year for *Red Dean*, but once work on this radar-guided missile commenced, things went nowhere fast, as will be seen. Finally, in 1954, a requirement specifying head-on or all-aspect engagement returned with OR.1131, code named *Red Hebe*. This 'Red' pair of missiles were related, with much of the work undertaken for *Red Dean* carried forward into *Red Hebe*, which was written specifically to arm OR F.115T, won by Fairey with the ER.103C.

If ever there was a project a company should have avoided a contract to build, it was *Red Dean* and Vickers. Originally begun by Folland in 1951, with the seeker head developed by E.K. Coles, the project very soon ran into trouble. After tackling increasing missile size and weight, but especially problems with the seeker head, Folland, already heavily involved with development of what would eventually become the Gnat, advised the Ministry of Supply that it could no longer continue the work and requested that the contract be cancelled. The Ministry of Supply then made enquiries with Vickers, which initially declined but agreed to continue the work following considerable pressure from the ministry, with a contract placed in 1953. Given the exacting nature of OR.1105, it is hardly surprising that Vickers was reluctant to take over the development, which had intricate instructions all the way down to the operational level, covering storage, assembly and maintenance of the missile by service personnel.

The design team may have changed, but the problems persisted, especially with the seeker head, which was now in the hands of G.E.C. It became something of a vicious circle; poor seeker performance meant using a larger warhead to destroy a target with a more distant near-miss, which meant more weight, which in turn necessitated a more powerful rocket motor, but that too led to even more weight. Improvements in materials technology also had to be sought as well; the lightweight aluminium initially chosen to reduce the weight was discovered to be unsuited for external carriage on supersonic aircraft, deforming in flight on its launch rail. A bonded honeycomb structure was eventually settled upon to cure the issue.

Red Dean launch trials were eventually carried out by Vickers utilising a Canberra, WD956, which provided just enough space for instrumentation and an observer. The trials usually took place over the Aberporth ranges off the Welsh coast, but later also at Woomera in Australia. By 1955, it was decided *Red Dean* should become the primary weapon of F.153D, the 'Thin Wing' Gloster Javelin. The marriage was to prove short-lived, unfortunately; the limited performance expected even with the more advanced, yet still

Close-up of the wing design showing the honeycomb filling used to reduce weight while maintaining stiffness.

***Above and below*: Vickers Type 888 – *Red Dean* sat on its display stand at the RAF Museum Midlands. At just over 16ft long and 12in in diameter, this 1,330lb missile is huge. When the official requirement specified carriage of two of these weapons, you can imagine the reaction of the aircraft designers. The follow on *Red Hebe*, was no better.**

subsonic, version of the Javelin, combined with the fact that government ministers had been shown the details and flight film of the vastly superior Canadian Avro CF-105 Arrow, resulted in the cancellation of the Javelin project in 1956. Work on *Red Dean* continued nonetheless, including plans for semi-recessed carriage of the missile under the fuselage of the CF-105,

A pair of *Red Dean* on display, one used to test the guidance system and the other red-nosed missile, the auto pilot. (Author's photography courtesy of Brooklands Museum)

as well as one option to create a fighter version of the Canberra, with a *Red Dean* on each wing tip.

Much of the effort expended by Vickers on *Red Dean* became the groundwork for *Red Hebe*, which was destined to arm the aircraft chosen as the winner from the proposals submitted for Operational Requirement F.155T. This was intended to become the main interceptor for UK air defence, expected to come into service in 1962 to take on high-flying supersonic Soviet bombers. From eight proposals submitted, two manufacturers were awarded development contracts; Fairey for its ER.103 Delta 3 (*see* Fairey Delta 2), with Armstrong-Whitworth's AW.169 as a back-up. In terms of weapons size, however, nothing had changed between *Red Dean* and *Red Hebe*; in fact, the weight issues were initially worse. At 17ft 6in long and weighing in at 1,300lb, *Red Dean* wasn't exactly lightweight, and *Red Hebe* would have been 22ft 7½in long with a similar weight, though the missile body had been slimmed by ½in. The arrival of reliable transistors to replace valves in some of the sub-systems within the missile would have finally reduced the overall weight of *Red Hebe*, but ultimately, any prospective UK purchase of CF-105, *Red Hebe*, or F.155T failed to survive the 1957 Defence White Paper.

Canada was left in similar shock when CF-105 was simultaneously cancelled, leading to a rapid southerly exodus of skilled aircraft-industry talent to well-paid jobs in the USA and effectively ending native Canadian advanced military aircraft projects for decades. The Canadian government did seem to understand better than its UK counterpart that manned aircraft could not be replaced solely by missiles, so bought the F-101 Voodoo from McDonnell Douglas for the RCAF, alongside the ramjet-powered Boeing CIM-10 Bomarc interceptor missile.

Vickers' model for its 1955 F.155T submission, the Type 559. One of the two pairs of missiles carried in the unusual dorsal launch positions would have been *Red Dean*, with the other missile possibly *Red Hebe* or *Firestreak*.

Chapter 14

Rotary Affairs

Britain appeared somewhat slow in realising the utility and versatility of rotary-wing aircraft, despite the presence of 12 Cierva C.30 autogiros in RAF service from 1934, operating for the School of Army Co-operation. There was very little obvious development during the war years, but behind the scenes, work was being carried out and once they were finally asked to produce new native designs, the creative minds of British industry applied a not inconsiderable amount of effort into producing something different. Once again, there are successes, but also more cases of 'so near, yet so far'.

Cierva as a company had carried on with autogiro development despite the death of founder Juan de la Cierva in an air crash at Croydon in 1936. Designs included true helicopters from 1938 with the W.5 and by 1948, the tri-rotor W.11 Air Horse heavy lifter. However, the prototype W.11, the largest

One of Britain's successful early helicopter designs, the Bristol Sycamore, here at Aerospace Bristol. Built at Filton in 1957, Sycamore HR.14 XL824 served with the RAF for a decade, finishing its days at the helicopter central flying school.

In the foreground at the Newark Air Museum is diminutive Saro Skeeter, XL764, minus its rotors, while looming behind is the much larger Westland Whirlwind HAS.7 XM685. The Whirlwind here is still piston-powered, with a 75hp Alvis Leonides engine replacing the original Wright Cyclone.

helicopter in the world at the time, crashed on 13 June 1950, killing test pilot Alan Marsh, Ministry of Supply test pilot John Cable and flight engineer Joseph K. Unsworth. As a result of this crash, Cierva's long-time financier, Scottish industrialist James G. Weir, withdrew further funding and soon afterwards, the company passed into the hands of Saunders-Roe. Work carried on with a Cierva design already on the drawing boards for what became the Saro Skeeter, a small two-seater training and scout helicopter that eventually entered service with the UK armed forces in 1956 as well as the West German army and navy.

The biggest problem facing all the early helicopters was a poor power-to-weight ratio, reliant as they were on piston engines. Although piston engines were powerful enough to get a conventional aircraft off the ground, when used to generate lift from rotating blades, their size and power-to-weight ratio limited

With Royal approval. Another Westland-built Sikorsky design, the Wessex, this is the former Royal Flight HCC.4 XV733, looking resplendent at the Helicopter Museum.

Wessex HC.2 XR525 at the RAF Museum Midlands, showing the twin Gnome engine installation. Adoption of a pair of de Havilland (later Rolls-Royce) Gnome turboshaft engines in the Wessex and later marks of the Whirlwind greatly improved their performance compared to the original piston engine. The Wessex enjoyed 30 years of service with both the RAF and Royal Navy.

the useful load that could be lifted. The solution adopted was to replace piston engines with smaller, more powerful gas turbines, allowing fitment in the same airframe space but gaining vastly better power-to-weight ratios in the process. In the UK, Westland carried this out with great success using Sikorsky designs built under licence. This began with the H-19 Chickasaw, which became the Westland Whirlwind, firstly switching the US-sourced Wasp or Cyclone radial piston engines for more powerful Alvis Leonides. Later came a Bristol Siddeley Gnome turboshaft to produce the Whirlwind HAR.9. This was followed by Sikorsky's H-34 Choctaw as the Wessex, initially using a Napier Gazelle gas turbine, but later the Bristol Siddeley Gnome once again, with both helicopters serving the RAF and Royal Navy exceptionally well for many years.

As opposed to a relatively simple turbine conversion, Fairey followed a totally different path to rotor blade power.

Jet Tips at Fairey – Ultralight and Rotordyne

Conventional piston-powered helicopters rotate their main rotor head using a shaft directly from or via a gearbox driven by the main engine. Power is applied uniformly across all blades, termed 'collective', while pitch control is achieved via a swash plate to adjust the rotor blade angle and hence the direction of thrust, which is termed 'cyclic' control. An additional driveshaft usually feeds a rear gearbox powering a tail rotor that provides contra-torque thrust and hence yaw (left/right) control, with the pilot using foot pedals to manage this axis.

Despite the success of the Fairey Delta 2 supersonic development aircraft, the last fixed-wing aircraft in service to bear the company's name was the turboprop Gannet anti-submarine and later AEW aircraft, of which, sadly, no prototypes remain. Prior to both these aircraft, in 1946 the company had started development of the Gyrodyne, a rather unusual rotorcraft that attempted to rethink the way helicopters worked. This machine would now be called a compound helicopter, having a mast-mounted rotor plus a pair of pusher propellers on stub wings. The main rotor and the propellers were all driven from a single Alvis Leonides 523 radial engine housed in the aircraft fuselage, with power delivered to the rotor shaft using a conventional Sikorsky-type swash plate driving the rotors, allowing helicopter-like operation. Once in horizontal flight, the drive to the main rotor could be cut, lessening the load on the rotor head. The aircraft then acted much like a normal autogiro, with forward propulsion from the wing-mounted propellers and the forward air speed maintaining the main rotor speed to produce lift. Two development aircraft were completed under a government contract to Specification E.4/46 and, after appearing in the static park of the 1947 SBAC Show at Farnborough, finally flew in December of the same year.

The Gyrodyne was a conventionally powered helicopter; consequently, half the machine's weight was attributed to the piston engine and drive mechanism. But confidence in the aircraft was such that in June 1948, the prototype, registered G-AIKF, set the first record for any helicopter in the UK and a world record of 124mph. Tragedy struck some ten months later, however, during an attempt at the 100km closed-circuit record, when the aircraft crashed near Reading, killing the pilot and his observer. The crash was eventually attributed to metal fatigue resulting in a failure within the rotor head. Following the investigation, the second prototype, G-AJJP, was grounded, but the story didn't end there; this aircraft re-emerged in a subtly modified form four years later as the Jet Gyrodyne.

The Jet Gyrodyne was a 'jet' only in how it powered the main rotor for take-off. Retaining the Alvis nine-cylinder piston engine of the original Gyrodyne and driving pusher propellers on each stub wing, it also gained a pair of superchargers from a Rolls-Royce Merlin engine. These were both driven by the Alvis engine but employed solely to produce compressed air. This air feed was piped to a combustion chamber mounted on the tips of the dual-blade

The Jet Gyrodyne at the Berkshire Museum of Aviation. Not until you get close do you realise just how long the twin-blade rotor is.

Starboard side view of the Jet Gyrodyne, showing one of the propellers for forward propulsion, both driven from the same Alvis Leonides engine.

rotor. Here, fuel was injected into the air stream and ignited, the resulting thrust powering the main rotor semi-independently of the pusher propellers and generating sufficient lift to allow take-off. Once airborne, the pusher propellers provided adequate thrust to allow the aircraft to transition from the hover to forward flight and, above a certain speed, the fuel feed to the tip jets could be cut. In all other respects, it behaved the same way as the Gyrodyne. The semi-independent power delivery to the rotor head also resulted in minimal torque being applied to the airframe and removed the need for a tail rotor to counteract it, both Gyrodyne variants having aircraft-like tail fins to provide yaw control as a result.

Development of rotor tip jet-powered helicopters now diverged to opposing ends of the aircraft spectrum, lightweight and heavyweight. Jet-powered rotor tips sounded good in theory; so much so that a much larger version, the Rotordyne, was planned, and Fairey was commissioned by the Ministry of Supply to use the Jet Gyrodyne as a research vehicle for the larger machine.

The Jet Gyrodyne first flew in January 1954, but transition from the hover to forward flight didn't appear until March 1955. Some of the delay was due to the aircraft being somewhat restricted by limited internal fuel capacity, resulting in a short flight duration and a lack of power, with the former often addressed by carriage of external fuel tanks under the wings. Once the prototype Rotordyne was delivered, the Jet Gyrodyne was retired and, although scheduled to be scrapped in 1961, it survived and is part of the RAF Museum collection. Today, it resides on loan at the Museum of Berkshire Aviation, not far from Fairey's former home airfield at White Waltham.

The Rotordyne was designed as a city-centre-to-city-centre passenger aircraft, to a requirement originally formulated in 1951 by British European Airways (BEA) initially for 30 passengers. The Ministry of Supply sponsored the design studies to meet this requirement, both agencies believing that this mode of transport could soon become the norm, and an official contract was signed with Fairey in April 1953 to build a single prototype, called Rotordyne Y.

With a proposed cruising speed of 150mph, a range of 250 miles and a passenger capacity now expected to be between 40 and 50, the Rotordyne would have been the largest transport helicopter of the day. Operating at low altitudes allowed the aircraft to be unpressurised, which also simplified construction; a square-section stressed-aluminium fuselage terminated in a tail section comprising clamshell doors for cargo operations, while short, high-set wings were carried plus a pair of large tailplanes on stub wings either side of the upper rear fuselage. With potential for both civil and military customers in the UK, as well as interest in the project from

Rear view of Fairey Rotordyne XE521, showing the clamshell cargo doors, tail assembly and tip jet unit on the nearest rotor.

US helicopter company Kaman to the extent of licensed production, plus enquiries arriving from airlines in Japan and Canada, the future looked bright; or so it appeared.

As with the Jet Gyrodyne, compressed air was generated by the engines and fed to rotor tips, where a fuel/air mix was ignited in combustion chambers to produce thrust. For the Rotordyne, however, Fairey had decided to use a pair of Napier Eland turboprops to power the aircraft, though the Rolls-Royce Dart and Armstrong Siddeley Mamba had been considered. It transpired that either of the latter duo might have been the better option, as Fairey later became highly dissatisfied with Eland development; Napier proved unable to maintain the rate of progress in improving the Eland that the project needed. Napier itself was subsequently taken over in 1961 by Rolls-Royce, which promptly axed the Eland.

The government had offered to provide funding for the project until the end of September 1957 but, despite the defence budget providing much of the funding to get this far, by the time of its first hovering flight on 6 November 1957, the Army and RAF had both lost interest in any military application. As a result, Rotordyne became a private venture, funded by Fairey and Napier. The first successful transition from vertical to forward flight and back to the hover finally came on 10 April 1958, with the development flying programme gaining pace over the following year, culminating in a record-breaking flight that gained the world speed record in the 'convertiplane' category over a 60-mile closed circuit of 190mph.

The only remains of the Rotordyne is the section of XE521's fuselage, alongside the rotor head and an example of the Napier Eland engine, all now on display at the Helicopter Museum in Weston-Super-Mare.

Rotordyne in flight. At this point, the tip jets would have been inactive, and the rotors would generate lift in the same manner as an autogyro.

After the amalgamation of Fairey into Westland Helicopters, the Rotordyne survived largely at the expense of Westland's own, more conventional transport helicopter design, the Westminster. With neither having any definite orders, but with the Rotodyne gaining more interest, company funds were allocated accordingly. By now, the Rotodyne prototype had conducted 120 hours' flying time over 350 flights, but storm clouds, predominantly of a financial nature, had begun to gather behind the scenes. The government mandated that further public funding would only be provided if definite orders were placed by BEA, which now wanted a larger aircraft capable of carrying 50 plus passengers. This would have meant a larger aircraft and consequently more powerful engines, with Fairey pencilling in Rolls-Royce Tynes for what would have been the Rotordyne Z; it never materialised.

Throughout the whole development programme, one major factor other than funding had been raising concerns, and delays in addressing this are a major reason that a production version never materialised: noise. For a machine intended to operate from city centres, the sound of two Eland turboprops was likely bad enough, but the noise level generated by the rotor tip pressure jets when ignited was said to be 'excessive', even though once up to flying speed, as with the earlier piston-driven aircraft, the pressure jets could be shut down. Despite a series of landings and take-offs being carried out from the Battersea Heliport with no registered complaints, it is perhaps the comment made by John Farley, who later went on to become chief test pilot for the Hawker Siddeley Harrier, that describes it best:

'From two miles away, it would stop a conversation. I mean, the noise of those little jets on the tips of the rotor was just indescribable. So, what have we got? The noisiest hovering vehicle the world has yet come up with and you're going to stick it in the middle of a city?'

Various attempts were made at noise reduction. An assortment of complex shaped jetpipe extensions were fitted, with appreciable effect, as a 10db noise reduction was eventually achieved. A completely new rotor blade was being designed that replaced the single-tip jets with a much longer pressure jet that would exhaust along virtually the whole trailing edge of the rotor blade itself, the combustion unit being built in the rotor blade aerofoil section, and it was hoped this would reduce the noise even further. It wasn't enough for BEA, which withdrew its support, effectively terminating the project.

The same fate that later befell TSR.2 was meted out to the Rotordyne, but this time with no survivors; the sole prototype was destroyed, and even the design documentation has also apparently disappeared. The meagre remains of the Rotordyne are on display at The Helicopter Museum in Weston-Super-Mare.

As mentioned, at the same time as Rotordyne was being developed, Westland was running a private development project called the Westminster, despite no official government requirement or funding, and had already spent around £1.3m by the time Fairey was incorporated into the company. It was intended to occupy a similar niche to the Rotordyne as a 30–40-seat civil passenger aircraft, though in addition, the skeletal underframe of a prototype often seen on displays was envisaged as a flying crane with a 15,000lb underslung load, double that of the Bristol Belvedere then in service with the RAF. With the upheaval of industry mergers in the late 1950s, and as the Rotordyne was still being funded by the government, the Westminster was abandoned.

The Westminster was not, as ever, a wholly Westland design, as it was based on the mechanicals, rotors and control systems for the Sikorsky S-56/CH-37 Mojave, but once again, like the S-55 and H-34, Westland replaced the original pair of Pratt & Whitney Double Wasp piston engines with British gas turbines, the same Napier Eland as the Rotordyne. The mechanical systems, which had only been loaned from Sikorsky, were returned to America (in sufficient time to avoid import tax!) and the two prototype airframes were scrapped.

It is arguable that the more conventional Westminster may have been a better bet for Westland than the Rotordyne, though it would have seen competition from early versions of the French Aerospatiale Super Frelon and Sikorsky CH-53 Sea Stallion.

Fairey Ultralight

Though most attention has focused on the Rotordyne, a considerably smaller Fairey-designed, jet-tip-powered machine was tested by both the British Army and the Royal Navy. This was the Fairey Ultralight.

It was built to what at first glance was a simple 1953 Minsitry of Supply requirement for a two-seat lightweight training, reconnaissance and casualty evacuation helicopter. This requirement had the caveat that the machine should be transportable on the back of a standard Army three-ton truck, which considerably restricted the dimensions, while the general performance specifications were particularly demanding. Fairey submitted the Ultralight to the Ministry of Supply, which was suitably impressed and awarded a four-aircraft development contract, against competition from Short, Bristol, Percival and others.

Powered by a compact Turbomeca Palouste gas turbine engine, which Sud Ouest had successfully used to power the equally diminutive tip jet-powered Djinn helicopter, the Ultralight was a simple design. A tubular box frame, with a Perspex-covered cockpit, basic skid landing gear and no need for tail rotors due to the low-torque tip jet drive, it met the Ministry of Supply demands

Westland's Westminster prototype G-APLE, here looking quite conventional with cladding hiding the tubular frame. A second prototype, G-APTX, was built with a fully clad fuselage and a six-blade rotor replacing the original five-blade to reduce vibration.

G-APLE, minus the cladding, operating as an aerial crane.

The Helicopter Museum's Fairey Ultralight G-AOUJ. The odd nose extension is to allow a stretcher and casualty to be carried inside the cabin.

to the letter, even being able to sit on the back of a standard army three-ton truck. However, it is doubtful that the Ministry intended an actual take-off or landing on the aforementioned three-tonner, but Fairey demonstrated this feat on several occasions just for good measure.

Despite some very successful trials and demonstrations, external factors terminated the Ultralight project before any more than the eventual six development machines had been produced. Firstly, the British Army favoured the more conventional Saro Skeeter for the role envisaged for the Ultralight. Secondly, West Germany advised that it would only buy the Skeeter if the UK did likewise, so the government ordered the Skeeter for the Army. Fairey also demonstrated the Ultralight to the Royal Navy as a shipborne helicopter for roles eventually undertaken by the Westland Wasp, with sea trials carried out on the relatively small 2,000-ton destroyer HMS *Granville*, but to no avail.

Despite overtures to the civilian market, which of course involved using company funds, development was abandoned in 1959.

Right: A close-up of the rear of the Ultralight cockpit, with the large tube up to the rotor head that delivered gas pressure to the rotor tips.

Below: Fairey Ultralight on trials with the British Army, seen here taking aboard a patient on a stretcher.

The Fairey Ultralight performs its 'landing on a 3-tonner' party piece.

Sea trials for the Ultralight in 1957 aboard HMS *Grenville*.

Bristol 173 and Belvedere

Bristol Aeroplane Company was one of the other early pioneers of British helicopter design; Bristol's Type 171, which became the Sycamore, enjoyed 20 years of service with the RAF, in addition to earning export orders for both military and civilian customers. The follow-on twin rotor Type 173 has rather cruelly been described as a pair of Sycamores bolted back-to-back, though look closely enough and it's not hard to work out why. Regardless, the 173 was the world's first twin-engine, twin-rotor helicopter.

Both the Sycamore and 173 were powered by Alvis Leonides 550hp radial piston engines, but the 173 housed a pair of them, with a common shaft drive allowing one engine to power both rotors in the event of an engine failure. Despite the 173 actually being 6ft shorter than the Sycamore, the new fuselage could carry 13 passengers, rather than the three in the single-rotor machine.

After its first flight in January 1952, initial teething problems with the 173 were ironed out, though later fuselage resonance problems with the original three-blade rotor resulted in this being replaced with a new four-blade design. Registered as XF785, the 173 had trials with both the RAF and the Royal Navy, which included sea trials off the aircraft carrier HMS *Eagle*. A second prototype, XH379, was built, fitted with forward and aft stub wings to improve cruising speed, but following more trials with the Navy, these were replaced with just a horizontal tailplane. Unfortunately, XH379 was written off following a crash at an airshow at Filton in 1956, though the tailplane was salvaged and fitted to XF785 to assist in solving continued stability issues.

The RAF had wanted a twin-rotor machine to lift underslung bulky loads, while the Navy seemed to be keen on ordering the aircraft for anti-submarine warfare, general transport and search and rescue duties, with the Canadian Navy also looking to adopt it for similar roles. The three orders, totalling 96 aircraft, would have been fulfilled by the Naval Type 191, RAF Type 192 and the Type 193 for the Canadians, all with Napier Gazelle turboshaft engines replacing the piston engines of the original Type 173 prototypes.

However, the Canadian order was cancelled to cut costs, and the Royal Navy opted to buy the Westland-built Sikorsky S-58, which became the Wessex, blaming a series of engine overheating problems during development flights with the third Type 173 prototype.

The RAF did ultimately adopt the Type 192 into service as the Belvedere; the first flew in July 1958 and was declared operational in 1961. The type saw service in Europe, Africa and extensively in Borneo with British forces during the Indonesia–Malaysia confrontation, where its heavy underslung load capability and the ability to carry 18 passengers was used to transport troops and equipment to jungle outposts.

By 1962, the RAF had also adopted the Wessex, and it started supplanting the Belvedere in many roles, though Bristol's machine was retained due to its outsize load-carrying ability. The Belvedere remained on strength until March 1969 when it was officially retired, its short service life in part a legacy of the design being based on the original Naval requirements, which called for under-fuselage anti-submarine torpedo mounts and hence the tall undercarriage. In the RAF transport version, this placed the cargo door 4ft off the ground once the undercarriage legs had been lengthened again to improve rotor clearance. With the engines being mounted at each end of the fuselage, this also prevented any redesign to incorporate a rear loading ramp, which would have been preferable. Belvedere's exit from service was also attributed to reliability issues and high maintenance costs, as well as a low fatigue life.

If the RAF or Army had wanted a replacement for Belvedere with a tail-loading ramp and better payloads, Westland had designed and offered the WG.11 in the mid-1960s. This quad Gnome-engined transport helicopter could have given the UK forces a Sea Knight/Chinook-class machine well before the real thing arrived in 1980.

Belvedere XG454, formerly housed at the Manchester Museum of Science and Industry, but now found at The Helicopter Museum following the closure of the Manchester site.

Bristol 173 G-AMJI, the second of the two prototypes built, with the rear wing extensions fitted.

***Above*: Now housed at Aerospace Bristol, minus rotors, is G-ALBN. Compared to the Type 192 Belvedere that followed, the undercarriage on the 173 is far better for passenger or cargo access but was raised on the Type 192 to improve rotor clearance during loading.**

***Right*: A pair of Belvedere HC.1s of 26 Squadron operating from RAF Khormaksar in Aden during the early 1960s. XG463 on the right crashed in 1964 following an engine failure.**

Stansing on the pan at Filton is Bristol 173 G-AMJI after gaining military serial XH379 for services testing.

Saunders-Roe P.531

As with many second-generation rotorcraft, the Saro P.531 was a turbine engine progression from an earlier piston-driven design, in this case the Skeeter. Started as a private venture in late 1957, the first P.531 flew in July 1958, powered by a 325shp Blackburn Turbomeca Turmo. It utilised many Skeeter components, including the short, wheeled undercarriage and rotor blades, but with a four-blade assembly in preference to the three blades on the Skeeter.

Following Westland's acquisition of the helicopter divisions of Saunders-Roe and Bristol in 1959, significant changes were made to the P.531 design, including doubling the engine power and replacing the wheeled undercarriage with skids. By now, an official request from the Ministry of Defence covered development costs, and this was for a five- or six-seat general-purpose helicopter for both the Army and the Navy. Several different engines were also tested during development, including the oft-used de Havilland Gnome and the Bristol Siddeley Nimbus, with the latter eventually chosen, though by the time Scout and Wasp production was complete, the Nimbus had become a Rolls-Royce product following this company's acquisition of Bristol Siddeley in 1966.

For the army, what became the Westland Scout retained a skid undercarriage, but the cabin was enlarged to provide space for five; the pilot and observer in the front and three rear passengers. Later models were fitted with bulged rear doors to allow a four-person rear bench seat, though more often, the doors were not fitted. In a casualty evacuation role, two stretcher cases could be accommodated either internally or in external pods. Provision was made for weapons carriage, and during its service the Scout often carried a single GPMG on a pintle mount with one door removed and was also cleared for firing four SS.11 air-to-surface guided missiles, carried in pairs either side of the fuselage on rail launchers, with a sight over the left seat on the cabin roof. Despite numerous trials over the years, no other weaponry was officially adopted.

The Scout entered service in 1961 and served worldwide with the British Army, participating alongside NATO forces in northern Europe, Northern Ireland, Borneo and in the Falklands War, finally leaving service in 1994.

The dedicated naval version was eventually named the Westland Wasp. The wheeled undercarriage of the Skeeter returned, but in a unique wide-track, four-legged assembly with castoring wheels to aid handling, plus lashing points to secure the aircraft to a wet rolling deck. Space for additional fuel tankage was found behind the cabin, which extended the range and hard points initially for anti-submarine torpedoes, though later the Wasp was equipped with air-to-surface guided weapons. Wasps from HMS *Plymouth* and *Endurance* successfully employed Nord Aviation AS.12 missiles during

Saro P.531-2 prototype at The Helicopter Museum. Much further advanced in the development process, it looks more like the finished product, which would be the Westland Scout in this case.

Despite bearing Royal Navy markings, this Saro P.531 prototype looks more akin to a Scout than a Wasp with its skid undercarriage.

Berkshire Museum of Aviation's Westland Scout XP489, in the colours of the Empire Test Pilots School.

the Falklands War to sufficiently disable Argentinian submarine *Santa Fe*, preventing it from submerging, this after it had been forced to surface by depth charges and anti-submarine torpedoes discharged from HMS *Antrim*'s Wessex and HMS *Brilliant*'s Lynx helicopters.

The Wasp enjoyed a 24-year career with the Royal Navy, giving even small ships an effective long range anti-submarine warfare capability, and its versatility was utilised in numerous other roles.

Both the Scout and the Wasp sold well overseas, to military or government buyers in South America, Africa, the Middle East, New Zealand and the Far East.

Two more Saro P.531 prototypes; these look much more like the eventual Wasp, with well-spaced wheeled landing gear.

***Above and below*: The Helicopter Museum's Westland Wasp XT433 last served on HMS *Aurora*. These were a regular fixture on RN vessels from their introduction in 1961 until replaced by the more capable Lynx in the late 1970s. The lower photo shows the folding tail rotor assembly, essential to allow stowage in the confined hangar space on smaller vessels.**

Chapter 15

Restoration

Regardless of when the aircraft was built, whether in wood and canvas or titanium and puddle-welded stainless steel, the aircraft in museums and those still in an airworthy condition are reliant upon the dedication and expertise of a host of people behind the scenes. Much as classic cars can need a continuous stream of time, money, love, attention and the occasional beating with a hammer, the same can be said for historic aircraft, but on a greater and more costly scale, while the engineering skills needed are not exactly readily available. This small army of engineers, many of whom are unpaid volunteers, are as dedicated in their task as those who designed and built the aircraft in the first instance and sadly, the passage of time will inexorably reduce this talent pool and knowledge base for older aircraft.

The biggest impact on the longevity of aircraft still intact but needing attention or those already restored is where they are kept, so those that sit outdoors in all weathers have the toughest life and need careful monitoring to check that water hasn't managed to seep in somewhere and started to eat away the aircraft from within. Be it good old rust, chemical or dissimilar metal corrosion and in some cases even woodworm, all need addressing as soon as possible to reduce the chances of long-term problems. For those aircraft stored under cover, the risks are greatly reduced, but humidity can still pose an issue, and should some furry rodent find access to a cosy home, wiring, leather or anything else chewable is in deep trouble.

For aircraft in various states of disrepair, or in some cases what looks like utter ruin, their road to recovery and display can be long and torturous. Witness the photographs in both volumes taken within the Fleet Air Arm's Cobham Hall storage facility, where your hapless author asked the obvious question 'go on then, what's all the bits of aluminium?' only to be told that they were numerous parts of several crashed Fairey Barracudas, recovered from far and wide, that would eventually be un-mangled, reassembled and turned into what will become the only survivor of over 2,500 Barracudas built. It's then that you realise the absolute dedication involved in preserving our aviation heritage for future generations and as a tribute to those who gave their lives for our liberty.

At the Fleet Air Arm's Cobham Hall storage facility, forget the Percival Sea Prince, nice though it is! The collection of seemingly random pieces of metal on the floor, along with a host of others that were scattered around everywhere the curators had space, will in time become a Fairey Barracuda.

Hawker Kestrel FGA.1 XS695, while undergoing restoration at the Sir Michael Beetham Conservation Centre (MBCC) at RAF Museum Midlands in 2012.

***Above and below*: By 2013, the airframe is now fully painted and with correct decals applied. The one-piece wing unit is lowered into place after the Pegasus engine is reinstalled.**

Special delivery! Wings Aviation Museum's Kestrel XS694 as it returned from the USA. Unbelievable as it may seem, the main fuselage has fared reasonably well over the 50 or so years since it departed the UK.

Britain's Hawker Kestrel community doubled in 2022, in the shape of XS694's return to UK shores; it is now being fully restored at The Wings Museum in West Sussex.

The restoration of XS694 is not simply preparing a 'shell' for display; every nut, bolt, panel, instrument and wiring loom need to be restored to a condition as original as possible. Numerous internal supports and airframe structures had to be repaired or fabricated due to corrosion, damage or just absence, while the cockpit instruments, switch gear and a canopy needed to be sourced from somewhere. The aviation community in general has made significant contributions to help restore this unique aircraft; Brooklands Museum has donated a spare P.1127 wing from XP984, and the RAF Museum

A labour of love. The nose section has been separated from the rest of the fuselage, while the blue frame nearby has since been modified to allow the whole nose to be rotated. (XS964 photographs courtesy Wings Aviation Museum)

has graciously donated possibly the only surplus Kestrel undercarriage in the world to help the team at Wings get XS694 back together.

Shipped to the USA and originally assigned to Eglin Air Force Base in Florida as a Hawker Siddeley Kestrel XV-6A, XS694, wearing US code 64-18268, was later transferred to NASA's Langley Research Centre. After suffering a crash on 27 August 1967 at NASA Wallops Island facility, Virginia, which rendered it beyond repair, it was stripped for spares and spent much of its life as an attraction at a paintball range.

With planning permission in place for a new museum at Dunsfold Aerodrome in Bedfordshire, XS694 will become a fitting tribute to where Hawker built every P.1127, Kestrel and Harrier.

Above and opposite above: Gloster Meteor NF.14 WS788, required over 20 years' work following an even longer period spent outside as a gate guardian and is still an ongoing project. It now looks in far better shape than in this 2022 photo and has been painted in 152 Squadron colours. (Author's photographs, courtesy Yorkshire Air Museum)

Opposite below: It's not just rain and sun that are the enemy of a good paint job; any car owner who cares for their vehicle will tell you that dive-bombers of the feathered variety can do the most serious damage to paintwork. Newark Air Museum's Vulcan B.2 XM594 undergoing a good clean to protect and preserve the aircraft, as well as inspect for any problems.

WORKING
AREA
NO ENTRY
WORKING
AREA
NO ENTRY

INSPIRED
Scaffolding Services Ltd

Across the Irish Sea, the Ulster Aviation Society has in its possession the only Short SB.4 built, which has for several years been enjoying a significant amount of TLC. As the photographs attest, the society's master joiner Peter Morrison, ably assisted by William Smyth, have had their work cut out, especially in creating a set of replica wings.

***Above*: Ulster Aviation Collection's Peter Morrison hard at work on the SB.4. (Photograph courtesy Alan Jarden)**

***Opposite above*: The complex and unusual wing shape of the Shorts SB.4 currently under renovation at the Ulster Aviation Collection in Belfast. (Images courtesy Mark J. Cairns)**

***Opposite below*: Some considerable time later, with the trailing edges yet to be completed and have linen coverings applied. It should be remembered that, unlike a conventional wing, the aero-isoclinic wing on the SB.4 has no ailerons or elevators on the trailing edge; control is affected by movement of the final fifth of each wing. (Image courtesy Stephen Riley)**

G-14-1
Sherpa

ULSTER AVIATION SOCIETY
Flying Into History
G-14-1
CRJ200
Sherpa

Above: Six months later, and with work on the tail boom completed (safely wrapped up and seen hanging above), space is now available for the main fuselage. This is a huge job; many of the badly rusted panels will need to be replaced, but this can only be done after hundreds of rivets and screws are removed. The observant will notice that there seems to be part of the fuselage side missing; this had been cut out to allow a children's slide to pass through the cabin!

Opposite above: Where do you start? Within the South Yorkshire Air Museum is the Yorkshire Helicopter Preservation Group; here they have Westland Dragonfly HAR.5 WN499 in need of a considerable number of hours of effort to restore it for display.

Opposite below: In parts, it's even worse than it looks and multiple panels will need to be fabricated to replace those where corrosion has truly taken hold. In the hangar behind, the lengthy Dragonfly tail boom and rotor assembly are in advanced stages of renovation.

Addendum To Vol.1

The business of displaying and housing aircraft in museums is subject to change for assorted reasons. Since the publication of Volume 1, the following aircraft covered in that volume have moved or are moving to new homes, whilst others are no longer on public display, hopefully only temporarily.

Prone Meteor F8 WK935 has been donated by the RAF Museum to Newark Air Museum and will doubtless sit next to the Reid Siegrist RS.4 Desford, the other prone-position trials aircraft already there. Newark thus has two unique Meteors, with WK935 joining the RB.108 V/STOL engine test aircraft, Meteor FR.9 VZ607. During 2025, Newark itself has seen the departure of Supermarine Swift FR5 WK277, which is now at the South Wales Aviation Museum, St Athan.

In terms of cancelled projects and what might have been, the biggest news is that after decades hidden away in the depths of the RAF Museum stores, the remains of what would have become the Hawker P.1121 have been sent to Brooklands Museum, and after some renovations are now displayed in its 'Factory' area. This display, possibly for the first time ever, should provide an inkling of the overall size of an aircraft which, if it had been completed, could well have equalled the F-4 Phantom in terms of performance and versatility, ten years before McDonnell's legendary machine.

The Short SB.5 WG768, along with its optional T-tail unit, crossed the Irish Sea and now resides with the Ulster Aviation Society at its Long Kesh, Lisburn base.

The last surviving short-fuselage DH Comet 1, G-APAS, left Cosford for the South Wales Aviation Museum at St Athan, but sadly will no longer be able to be displayed with its wings attached, as extensive corrosion was discovered whist it was prepared for transit, and the cost of correction was determined to be prohibitive.

Both the Short SR.53 'Rocket Fighter' and the Bristol Type 188 are awaiting new owners, as the RAF Museum has opted to dispose of them. The latter may end up at Aerospace Bristol, housed on the site at which it was built, whilst the former, as of April 2026, is yet to find any takers.

Afterword

With the aircraft all being in museums, you'd think trying to piece together the contents of what turned into a pair of books would be easy enough. But trying to work out if a prototype for an aircraft that may have entered service nearly 70 years ago actually still exists is harder than you'd expect, even in the world of internet search engines. If an aircraft turns out not to be extant, then there's the matter of confirming what happened to it, all the while expecting someone to say 'oh, it's actually here', or 'there's one of those at such and such a place'.

Too late in the day for Volume One did I realise that I'd missed the third prototype Folland Gnat, when I'd been only a few miles away on a job for work. Hence the Gnat's inclusion in this volume alongside the Jet Provost, despite Folland becoming part of Hawker Siddeley in 1959 as opposed to BAC.

What started out as just adding words during the COVID lockdowns to an assortment of my photos I'd accumulated, just to keep me sane, evolved after thinking that it might look nice if I got it printed as a book just for me (having previously created photo books for a couple of weddings). It took on a working title of 'All That Remains', with my photographs showing what was left of the once thriving British aircraft industry of the immediate postwar years, the so-called 'Golden Age'. This morphed into two proper books after I replied to an advert in *Aeroplane Monthly* and the lovely people at Key Publishing decided my work was good enough for publication.

It was only then that I realised that it was even harder trying to get to some of the locations where the more obscure machines were to be found, especially in the case of those hidden away in storage from the main museums. Those were invariably the ones I'd neglected to photograph long before actually imagining writing a book, even though I'd had the chance. All on top of holding down a day job.

I didn't help myself when, not unlike several of the aircraft requirements mentioned, the contents within the two volumes became subject to 'scope creep', the scourge of any software developer or writer! Consequently, I ended up adding elements on subjects I'd not originally considered, such as Britain's nuclear weapons and or air-to-air missiles, but these were not only relevant to the overall theme but highlighted the difficulties the aircraft designers faced when 'the Bomb', the weapon their aircraft was being designed and built to carry, was still a work in progress itself.

Some of my photographs also date back to when a number of aircraft were previously displayed in other museums; the Avro 707C, for example, had disappeared from the Test Flight Hall at Cosford only a year after I took my photos, but now, most appropriately, has a new home at Avro's heritage centre near Manchester. Its 707A stablemate also moved after my visit to the Manchester Museum of Science and Technology; after the closure of the Air and Space Hall there due to the dire condition of the building, this now resides with the Boscombe Down Aviation Collection.

The seed for this pair of books was probably planted way back in the mid-1970s, when as a teenager I chanced upon Derek Wood's *Project Cancelled* in my local library and renewed my borrowing of it for months on end, unable to believe that the country could have built, let alone have in service some

Saunders-Roe SR.53 at the RAF Museum Midlands. Constant care means that it looks immaculate. As of early 2026, this and several other research and development aircraft are awaiting new homes, after the museum elected to dispose of those types that never entered RAF service.

of these amazing aircraft and how much money had been thrown away on developing them before they were cancelled. More recent works such as Damien Burke's excellent *TSR.2: Britain's Lost Bomber* (as well as his superb website (www.thunder-and-lightnings.co.uk) along with a good deal of Tony Buttler's published work, rekindled my interest and set me on the path to actually going out and seeing the aircraft that still survive.

Of the projects Britain cancelled and the opportunities that were missed, it's not hard to think that numerous governments and their respective swarms of civil servants could have done a lot better for both the armed forces and the country, regardless of financial resources. Those at the top of the armed forces didn't exactly help, especially where funding for one project meant less for another where a rival service was involved. The problems of the UK aircraft industry in the 1950s largely reside in the outlook of the governments of the time, regardless of their politics. Empire was at an end, Britain was no longer the great power it had been and trying to maintain what couldn't be afforded in projecting power onto foreign shores should have been recognised sooner, to allow concentration of resources on what was needed closer to home. The,

problems, however didn't sit at the door of the design departments of Hawker or Vickers et al; they delivered what was asked of them and occasionally produced aircraft which became icons. Where the aircraft themselves are concerned, hindsight is a wonderful thing, but there are simply too many 'what if' situations and all we can do now is dream of what could have been.

In completing this project and looking into the development of all the aircraft, I have felt fortunate and honoured to have had the chance to work on the systems for one of the eventual success stories, having served as an RAF aircraft electrician on the Hawker Siddeley Harrier. I departed as the plane that evolved in the analogue 1960s was superseded by the digital age and the largely American-redesigned version of the late 1980s. Its retirement in 2010, way ahead of the arrival of the replacement F-35B deprived the Royal Navy of fixed-wing combat aircraft for over a decade and the RAF of probably its most versatile asset, which to me says that politicians have learned very little since the 'Golden Age'.

Appendix

Visiting Aircraft

Former 111 Squadron Lightning F6, XS903. (Author's photograph, courtesy Yorkshire Air Museum)

Aircraft	Serial	Museum	Notes	
		RAF Museum Midlands, RAF Cosford		www.rafmuseum.org.uk
Bristol Type 188	XF926			
Hunting H.126	XN714			
Hunting Percival P.84 Jet Provost	XD674			
EE P.1A	WG760			
EE P.1B	XG337			

Aircraft	Serial	Museum	Notes	
BAC TSR-2	XR220			
Fairey Delta 2	WG777			
SR.53	XD154			
SR.177 Model				
Westland Wessex HC.2	XR525			
Red Dean Missile				
Hawker Kestrel	XS695			
		Norfolk and Suffolk Aviation Museum		www.aviationmuseum.net
EE F.1 / DB	XG329			DB – Development Batch
Jet Provost T3A	XN500			
TSR-2 part	?			
		Boscombe Down Aviation Collection		www.boscombedownaviationcollection.co.uk
EE P.1A	WG763			
		Ulster Aviation Collection		www.ulsteraviationsociety.org
Shorts SB.4	G-14-1			
Shorts SB.5	WG768			
		Berkshire Museum of Aviation		
Fairey Jet Gyrodyne	XJ389			
Westland Scout	XP489			
		IWM Duxford		www.iwm.org.uk/visits/iwm-duxford
BAC TSR-2	XR222			
		Brooklands Museum, Surrey		www.brooklandsmuseum.com
BAC TSR-2			Thermal test cockpit	

Aircraft	Serial	Museum		
Supermarine Swift F.4	WK198		Fuselage only	
Red Dean Missile				
Concorde	G-BBDG			
P.1121			Incomplete	
		FAA Museum, RNAS Yeovilton		www.fleetairarm.com
Handley Page HP.115	XP841			
Supermarine Type 510	VV106			Not on public display
BAC 221	WG774			
Saro P.531	XN332		Cobham Hall	Not on public display
Saro P.531	XN334		Cobham Hall	Not on public display
		Newark Air Museum		www.newarkairmuseum.org
Saro Skeeter	XL764			
Jet Provost T.3	XM383			
Westland Whirlwind	XM685			
Avro Vulcan B.2	XM594			
		Midland Air Museum		www.midlandairmuseum.co.uk
Fairey Ultralight	G-APJJ			
Folland Gnat F.1	XK741			
		Solent Sky, Southampton		www.solentsky.org
Saunders-Roe SR 1/A	TG263			
Supermarine Swift F.7	XF114		Under restoration	
Supermarine 545 model				

Aircraft	Serial	Museum		
		Aerospace Bristol		aerospacebristol.org
Bristol 173	G-ALBN		Fuselage only	XF785
Bristol Sycamore	XL824			
		The Helicopter Museum		
Saro P.531	XP165			
Bristol 193 Belvedere	XG454			
Fairey Rotodyne	XE521		Small section	
Fairey Ultralight	G-AOUJ			
Saro P.531-2	XP165			
Westland Wasp	XT443			
Westland Widgeon	G-AOZE			
Westland Wessex HCC.4	XV733			
Piasecki H-21			Under restoration	
		Wings Aviation Museum		www.wingsmuseum.co.uk
Hawker Kestrel	XS694		Under restoration	
		Yorkshire Air Museum		yorkshireairmuseum.org
Gloster Meteor NF14	WS788		Under restoration	
Dassault Mirage IV	45 'BR'			
Dassault Mirage III	538			

Aircraft	Serial	Museum		
		South Yorkshire Air Museum / Yorkshire Helicopter Preservation Group		www.southyorkshireaircraftmuseum.org.uk
Westland Dragonfly HAR.5	WN499		Under restoration	
		Hamble Aerostructures, Folland Sports and Social Club		
Folland Gnat T.1 Prototype	XM693		On display outside	
		South Wales Aviation Museum		
Supermarine Swift FR.5	XV277			

Bibliography

Books

Project Cancelled, Derek Wood, Jane's, 1975 and 1986
British Secret Projects: Jet Bombers Since 1949, Tony Buttler, Midland Publishing, 2003
British Secret Projects: Jet Fighters Since 1950, Tony Buttler, Midland Publishing, 2000
Cold War Interceptor, Dan Sharp, Mortons, 2019
Hawker's Early Jets, Christopher Budgen, Air World, 2022
Hawker P.1103 and P.1121, Paul Martell-Mead and Barry Hygate, Blue Envoy Press, 2015
TSR.2: Britain's Lost Bomber, Damian Burke, Crowood, 2010
BAC TSR.2: Lost Tomorrows of An Eagle, Paul Lucas and Andy Evans, SAM Publications, 2010
Vulcan's Hammer, Chris Gibson, Hikoki Publications, 2011
English Electric Lightning: Genesis and Projects, Tony Wilson, Mortons, 2021
From Jet Provost To Strikemaster, David Watkins, Grub Street, 2017
From Lysander to Lightning: Teddy Petter, Aircraft Designer, Glyn Davies, The History Press, 2014
The 'Secret' World of Vickers Guided Weapons, John Forbat, The History Press, 2012
TSR.2: Precision Attack to Tornado, John Forbat, The History Press, 2006
Swift Justice: The Full Story of the Supermarine Swift, Nigel Walpole, Pen and Sword, 2004
Fairey Rotordyne, David Gibbings, The History Press, 2009
From Sea to Air; The Heritage of Sam Saunders, A.E. Tagg & P.L. Wheeler, Crossprint, 1989
RAF Nuclear Deterrent Forces, Humphrey Wynn, The Stationery Office, 1994
Wrecks and Relics 26th Edition, Ken Ellis, Crecy, 2018

Bookazines and Magazines

Aviation Archive: British X-Planes – The Jet Era, Key Publishing
Aeroplane Monthly, Key Publishing

Acknowledgements

Many thanks to the good people at the Ulster Aviation Society for providing the photos of the Shorts SB.4 during the lengthy renovation work on the aircraft.

Thanks Aaron Simmons and the team at Wings Aviation Museum (www.wingsmuseum.co.uk) who provided the photos and information for Kestrel XS694.

Alan Beattie at the South Yorkshire Aircraft Museum / Yorkshire Helicopter Preservation Group for intimate access and updates on progress with the Westland Dragonfly and information on the Saro P.531-2.

Naturally, once again my appreciation of my family for the encouragement to finally finish the project. Trouble is, now they're worried I might find something else to go photograph and write a book on!